D0934046

EVOLUTION

THE DARWIN COLLEGE LECTURES

EVOLUTION

SOCIETY, SCIENCE AND THE UNIVERSE

Edited by *A. C. Fabian*

CAMBRIDGE
UNIVERSITY PRESS

PUBLISHED BY THE PRESS SYNDICATE OF THE UNIVERSITY OF CAMBRIDGE

The Pitt Building, Trumpington Street, Cambridge CB2 1RP, United Kingdom

CAMBRIDGE UNIVERSITY PRESS

The Edinburgh Building, Cambridge CB2 2RU, United Kindgom

40 West 20th Street, New York, NY 10011-4211, USA

10 Stamford Road, Oakleigh, Melbourne 3166, Australia

© Darwin College, Cambridge 1998

This book is in copyright. Subject to statutory exception
and to the provisions of relevant collective licensing agreements,
no reproduction of any part may take place without
the written permission of Cambridge University Press.

First published 1998

Printed in the United Kingdom at the University Press, Cambridge

Typeset in Linotron 300 Iridium 10/14pt

A catalogue record for this book is available from the British Library

Library of Congress Cataloguing in Publication data

Evolution: Society, Science and the Universe / edited by A.C. Fabian.
      p.   cm. – (The Darwin College lectures)
   Includes index.
   Contents: On transmuting Boyle's law to Darwin's revolution /
Stephen Jay Gould – The evolution of cellular development / Lewis
Wolpert – The evolution of guns and germs / Jared Diamond – The
evolution of London / Richard Rogers – The evolution of society /
Tim Ingold – The evolution of the novel / Gillian Beer – The
evolution of science / Freeman Dyson – The evolution of the
universe / Martin Rees.
   ISBN 0 521 57208 8 (hardcover)
   1. Evolution.   I. Fabian, A. C., 1948–   .   II. Series.
B818.E82   1998
116–dc21   97–23260 CIP

Contents

LIBRARY
ALMA COLLEGE
ALMA, MICHIGAN

Introduction

ANDREW C. FABIAN

The concept of evolution means different things to different people. To a biologist it simply means genetic evolution, whereas in many other disciplines it can mean change, or unfolding, with time – sometimes with an implicit gradualness to distinguish it from revolution.

This collection of essays is the result of asking eight well-known communicators from separate disciplines to discuss evolution. It will be seen that most of them tell us how the topic has arrived where it has: the Darwinian concept of evolution itself from Stephen Jay Gould; cells and the embryo from Lewis Wolpert; the current human political divide (with a very broad brush) from Jared Diamond; society from Tim Ingold; the universe from Martin Rees; the scientific enterprise from Freeman Dyson; Richard Rogers on the evolution of cities concentrates on the current state of London and Gillian Beer considers whether novels have evolved at all and then tells us how the concept of evolution has permeated through fiction.

The essays were originally given as the Tenth Darwin College Lecture Series in early 1995. These Series have become an institution for some in Cambridge and are open to, and well attended by, the public as well as members of the College and University. The Lectures are intended to be interdisciplinary and to that end we select for each Series a group of well-known communicators, mostly from the academic world, and ask them to talk on a chosen theme. We have now had Origins, the Fragile Environment, Predictions, Communicating, Intelligence, Catastrophes, Colour and now Evolution. A greater appreciation of the overall result can best be obtained by looking at more than one volume. Well-known names who have contributed include Noam Chomsky, Stephen Hawking, Roger Penrose, John Maynard Smith, Desmond Tutu, Helena Kennedy, Robert May, Richard Gregory, Daniel Dennett, David Lodge, Roy Porter, Bridget Riley and Christopher Zeeman.

There is no intention that the authors should know what the others have said or written. In this way the feel for each discipline comes through. Styles of writing vary as much as styles of lecturing. Generally, the sciences have more visual aids – illustrations and figures – and are more impromptu than talks from the arts, social sciences and humanities. By giving the talks to a wide public audience it is generally true that they are wholly intelligible to those outside their discipline. We hope that some of the excitement of the intellectual endeavours in a range of disciplines can thereby be readily communicated.

Although the process of Darwinian evolution is contested little in the chapters, recent interpretations (neo- and ultra-Darwinism) are questioned in places, and a suspicion that there is more to life grows as attention shifts from the physical and biological sciences to the social ones.

Stephen Jay Gould argues that the mechanism of evolution through adaptation has deep roots in the tradition of English natural history. This tradition, in which Darwin worked, emphasized the study of details and of good design and can be seen centuries earlier in the writings of Robert Boyle.

Lewis Wolpert shows that evolution proceeds by (genetic) modification of the development of the embryo, with differential growth being the important factor. He then explores how multicellular creatures, eggs and embryos arose.

Jared Diamond tackles the broad pattern of human history since the last Ice Age. Why did Europeans spread through North America after the fifteenth century whereas the Incas did not (and could not) invade Europe? He argues that much has to do with the availability of domesticatable large animals and plants which, in turn, depends to a significant degree upon geography. If a land mass (Eurasia) has a large east–west axis at the same latitude, plants and animals can be successfully transplanted. Large human societies, and farming, can develop. Surplus people can develop guns; germs are caught from the domesticated animals. The invaders then have weapons, transport and disease immunity so are unstoppable.

Richard Rogers considers that the evolution of a city such as London is a process of cumulative change which cannot be left to random mutation. London is a postindustrial city facing the consequences of unchecked economic growth. Many famous public spaces are little more than traffic roundabouts. He argues that a sustainable city can be developed by consolidation around diverse compact urban neighbourhoods.

Tim Ingold thinks that developmental biology, rather than Darwin's 'descent with modification' is the most promising place to start integrating the biological and social sciences. He argues against social life being a product of a selection process and that human capacities are not pre-specified but arise through evolution with other persons, particularly in the interaction with succeeding generations.

Gillian Beer finds that evolution as applied to the novel generally means upward development, and argues against the novel as such evolving. It is the case that novels that are fit to survive do so, and not necessarily those that have survived being somehow intrinsically the best. She then explores how evolution as an idea has been taken up by novelists.

Freeman Dyson tackles the evolution of science by telling stories, mainly of astronomy, to illustrate evolutionary themes. He shows how speciation and symbiosis occur in the physical universe. It is well known that revolutions occur in the development of scientific understanding. He argues that tool-driven revolutions are much more common than the, more famous, concept-driven ones.

Martin Rees describes the evolution of the universe as an unfolding process. This takes us from the Big Bang to the build-up of elements heavier than hydrogen and helium in stellar cores and supernovae. He then asks why the physical constants, the strengths of interactions and so on, have the values they do, and discusses ideas in which there are ensembles of universes of which ours is just one in which the constants allow life to develop and exist. This takes us full circle and allows us to glimpse how genetic evolution may be just a later part of a more directly physical, and possibly deeper, process of unfolding in a physical metauniverse.

We see that, in an interdisciplinary sense, the concept of evolution is not fixed, but is a central theme which cannot be ignored.

Lastly, I am grateful to the Master and Fellows of Darwin College for the opportunity to organize the Lecture Series, and to many individual Fellows and students for assistance in running the Lectures and, in particular, to Joyce Graham for much practical help.

1 On Transmuting Boyle's Law to Darwin's Revolution

STEPHEN JAY GOULD

Adaptive continuity

Perhaps there will not always be an England (particularly on time scales favoured by palaeontologists), but a few miles of Channel and nearly a thousand years of freedom from full-scale invasion (1066 and all that) have produced a plethora of British distinctions, both idiosyncratic and deeply philosophical, from continental preferences and modes of thought. (A common language across 3000 miles of ocean can inspire more closeness than twenty miles of *La Manche* accompanied by a divergence of tongues – hence the similarities between American and British histories of evolutionary thought, as discussed in this article.) In this work, I try to identify adaptation as the most distinctly anglophonic subject of natural history and subsequent evolutionary ideas. I set out to show that Charles Darwin's (Figure 1) decision to site his defence and mechanism of evolution in the explanation of adaptation has roots in a long tradition of English natural history and theology that never provoked much continental attention. Our current struggles over 'ultra-Darwinian' versus structuralist modes of thought continue the same debate and establish a particularly English continuity across several centuries.

In the operative paragraph of his Introduction to *The Origin of Species*, Charles Darwin stated (1859, p. 3) that the classical subjects of natural history could provide sufficient evidence for the factuality of evolution:

> In considering the Origin of Species, it is quite conceivable that a naturalist, reflecting on the mutual affinities of organic beings, on their embryological relations, their geographic distribution, geological succession, and other such facts, might come to the conclusion that each species had not been independently created, but had descended, like varieties, from other species.

FIGURE 1 Charles Darwin, by Leonard Darwin (1878).

He then added, in a portentous line that has sounded throughout the subsequent history of evolutionary theory, that such an explanation would seem
empty, not only for leaving out a central subject, but on aesthetic grounds as
well:

> Nevertheless, such a conclusion, even if well founded, would be unsatisfactory,
> until it could be shown how the innumerable species inhabiting this world have

been modified, so as to acquire that perfection of structure and coadaptation which most justly excites our admiration.

(Ibid.)

Darwin then cites his reasons for locating the causes of evolutionary change – not just its factuality, which can be otherwise ascertained – in both the *complexities* and *precision* of good organic design (and not just the simple existence thereof). Darwin invites us to consider the alternatives: how else, other than by natural selection, might precise adaptation arise by material causation rather than direct supernatural construction? Darwin notes that environmental induction of variation would be cited by most evolutionarily inclined naturalists, but such an explanation cannot account for the complexity and beauty of adaptation (an argument with a strong aesthetic component):

> Naturalists continually refer to external conditions, such as climate, food, *etc.*, as the only possible cause of variation. In one very limited sense, as we shall hereafter see, this may be true; but it is preposterous to attribute to mere external conditions, the structure, for instance, of the woodpecker, with its feet, tail, beak, and tongue, so admirably adapted to catch insects under the bark of trees.
>
> *(Ibid.)*

Add the Lamarckian notion of use and disuse (which Darwin labels 'habit') or direct organic will (his standard misreading of Lamarck, taken in part from Charles Lyell's summary in volume two of the *Principles of Geology* (1832)), and one might edge closer to an explanation for precision, but not for intricate coadaptation between ecologically interdependent organisms. Darwin continues (1859, p. 3):

> In the case of the misseltoe, which draws its nourishment from certain trees, which has seeds that must be transported by certain birds, and which has flowers with separate sexes absolutely requiring the agency of certain insects to bring pollen from one flower to the other, it is equally preposterous to account for the structure of this parasite, with its relations to several distinct organic beings, by the effects of external conditions, or of habit, or of the volition of the plant itself.

We are left with only one alternative to natural selection: the orthogenetic notion of a 'pre-programmed' sequence of phylogenetic transformation, as the Scottish author and publisher Robert Chambers advocated in the anonymously published *Vestiges of the Natural History of Creation* (1844). Darwin properly rejects this notion on methodological grounds – as entirely untestable in the same sense that creation by divine fiat can never be proven and cannot therefore be regarded as useful:

> The author of the 'Vestiges of Creation' would, I presume, say that, after a certain unknown number of generations, some bird had given birth to a woodpecker, and some plant to the misseltoe, and that these had been produced perfect as we now see them; but this assumption seems to me to be no explanation, for it leaves the cause of the coadaptations of organic beings to each other and to their physical conditions of life, untouched and unexplained.
>
> (Darwin, 1859, p. 4)

English-speaking evolutionists are so accustomed to accepting the primacy of adaptation that they tend to regard such paramountcy as self-evident and not subject to alternative construction. But a decision to view adaptation as the central phenomenon for evolution to explain represents a peculiarly English strategy, and by no means a universal approach. Darwin's revolution may be defined by its radically new and utterly inverted explanation of adaptation, but not by a decision to make the subject central – for good design had been the primary subject of English natural history for at least 200 years.

These differences in national styles, since they began long before the acceptance of evolutionary perspectives, arose from varying approaches to the question of how the workings of nature might reflect the presence and attributes of a divine creator. The distinctively English tradition of 'natural theology' held that God's existence, and also his attributes of goodness and omniscience, could be inferred from the excellence of organic architecture, particularly the good design of organisms and the harmony of ecosystems. Natural theology was defended by some of the greatest seventeenth-century scientists in Newton's orbit, Robert Boyle and John Ray in particular; achieved a culminating statement in the immensely influential *Natural Theology* of William Paley, first published in 1802; and enjoyed a final exuberant fling, a bit past its time perhaps, in the sequence of Bridgewater Treatises published during the 1830s. The natural theologians therefore viewed 'adaptation' – their word, by the way, not Darwin's invention or evolution's neologism – as the primary phenomenon of biology because God's existence and nature lay best revealed therein.

Such an attitude would have seemed peculiar to most continental biologists who did not (of course) deny adaptation, but who tended to view good design as a set of superficial and particularistic tinkerings upon the basic illustrations of divine intelligence: underlying structures, and the patterns of their trans-formation in the taxonomic ordering of animals. Most continental structur-alists viewed a well-webbed duck's foot, or a good-digging mole's forearm, as too singular and too puny to illustrate something so ineffable and general as God's omniscience. Louis Agassiz, for example, the great Swiss (and, later, American) zoologist of Darwin's generation, and the last major scientific creationist, held that the taxonomic structure of the animal kingdom best revealed God's nature and intentionality – for each species is an incarnated thought in God's mind, and relations among species therefore display the character of God's mental machinery.

I do not mean to cast this distinction as a pure and invariant dichotomy. Some continentals, notably the French naturalist Georges Cuvier himself, maintained a predominantly adaptationist outlook (non-evolutionary, of course, for Cuvier). And some Englishmen favoured the search for geometric rules of archetypal transformation over a singular focus on adaptations, each separately fitted to a particular environment – including Richard Owen, whose adherence to this unfamiliar style of evolutionism led to his frequent misinter-pretation (abetted by a growing Darwinian establishment, quite content to malign their principal enemy) as a lingering creationist (for non-adaptational evolutionism might easily be misread as a denial of the entire theme, rather than only of centrality for Darwin's favoured phenomenon). Paleyan natural theology may have been more the preserve of dons and divines in Cambridge than of the medical radicals in Edinburgh and London (who, as the biographer and historian of science Adrian Desmond has shown so well, often embraced Lamarckian and structuralist views); but Darwin ran with the Cambridge crowd, and this strand of intellectual genealogy ultimately prevailed in British biology.

I therefore consider it useful to examine the distinctively British continuity between the adaptationism of the natural theologians and its transmogrifi-cation into Darwin's world of descent with modification. The contrast has often been drawn between Paley and Darwin – and fairly enough, for the essence of Darwin's revolution may be defined by the causal inversion thus introduced – but few have focused on the equally striking continuity. Darwin, in short, kept the phenomenology and inverted the explanation – and we need to understand the part retained as well as the portion overturned.

Natural Theology has usually been characterized by its late and canonical expression in Paley (or by its death throes in the later Bridgewater Treatises). I would rather focus on the founding documents of Newton's age – particularly on my favourite work by the greatest of Newton's contemporaries who treated the subject explicitly and at length – Robert Boyle, in his 1688 work entitled *A Disquisition About the Final Causes of Natural Things, Wherein it is Inquir'd Whether, and (If at All) With What Cautions, a Naturalist Should Admit Them*. I want to examine how Boyle sets up the argument for organic adaptation as the primary natural clue to God's existence and attributes. I shall then discuss the features of his system that persist with most continuity into later Darwinian traditions, and also the components most radically overturned by evolutionism. In tracing this unbroken thread, I believe that we can also best understand the differences. As for the lineages of organisms that he studied in nature, Darwin's theory emerged in genealogical continuity with a local intellectual ancestry. We will best understand the truly revolutionary aspects of natural selection when we can map its explanatory inversion upon the unaltered conviction that adaptation represents the central phenomenon requiring explanation by any adequate theory of life's history.

Boyle's formulation

The architects of the scientific revolution (that is, the late seventeenth-century formulation of modern science that historians tend to dignify, often in upper case, as *The* Scientific Revolution) held a distinctive attitude towards the role of God in nature. All were devout theists, perhaps no one more so (or at least in a seriously conventional manner, for Newton certainly had maximal zeal) than Robert Boyle (Figure 2). They did not deny to God his traditional prerogative of miraculous interference into the affairs of nature, whenever he so desired or felt the need. Boyle, for example, writes in his *Disquisition* (1688, p. 96):

> Nor is this doctrine inconsistent with the belief of any true miracle; for it supposes the ordinary and settled course of nature to be maintained, without at all denying, that the most free and powerful author of nature is able, whenever he thinks fit, to suspend, alter, or contradict those laws of motion, which he alone at first established, and which need his perpetual concourse to be upheld.

But in general, and effectively all the time, God will not so intervene. A deity who must perpetually put his finger into nature's affairs, to correct some glitch that his own omniscience should have foreseen, is a poor and bumbling power

FIGURE 2 Robert Boyle.

indeed. How much more majestic to posit an infallible God who ordains all laws at the inception of the universe to produce the desired effects throughout later history and without further direct maintenance. 'This doctrine' (to use Boyle's words) of a 'clockwinder' God, who got the laws right at the beginning and thereafter let nature run by the invariant principles that he had ordained, forged a beautiful harmony between serious belief and untrammelled science – for God, as a perfect mechanic, combines maximal majesty with minimal perturbation. In short, the author of nature had made a world that science might fully comprehend.

But such an operational attitude entails a paradox. If 'the heavens declare the glory of God; and the firmament showeth his handiwork' (recall Haydn's setting of these words in his oratorio *The Creation* (1798) when you envisage the emotional power of this claim), then how shall we know this most fundamental of all truths? If nature now operates by invariant laws, where is God's imprint upon the works of his creation? No doubt he ordained the inverse-square law, but such mathematical abstractions seem a bit distant from our need to affirm his benevolence and his loving kindness towards humanity, the crown of his creation. How shall we know our favoured status? How shall we remain sure that 'though after my skin worms destroy this body, yet in my flesh shall I see God' (and now think of Handel's setting in *Messiah* (1741)).

The most attractive resolution of this paradox lay in the old Aristotelian doctrine of final cause. (Remember that Aristotle, in the *Organon*, divided causality into four distinct modalities, which he named material, efficient, formal and final. Using the familiar 'parable of the house', the standard pedagogical device for explicating this notion, material causes are the stuff of construction – straw, sticks and bricks offering different degrees of protection against wolves, for example. Efficient causes are the actual 'hands-on' makers of the effect – the mason who lays the bricks. Formal causes are the abstract plans or archetypes that govern the construction; blueprints do not make anything directly, but you will not progress beyond a pile of bricks without such a planned design. Final causes are purposes, for the house will not be built unless someone wants to live there and can commission the builders for this end.)

The scientific revolution placed such primacy upon efficient causes that modern usage has restricted the entire concept to only one of Aristotle's four modalities. We still acknowledge the importance of material and formal factors, but we no longer refer to them as causes. Final causes have been banned for inorganic objects (the moon does not exist to illuminate the night sky), and accepted as an unintentional consequence of natural selection in the evolution of organisms (moles do have stout forearms for digging, but they did not consciously strive to evolve such structures). The human brain grants us intentionality and final cause in the original sense, but we are in oddity in nature.

Final cause, however, remained a legitimate notion for scientists of Boyle's generation (despite Francis Bacon's famous deprecation). For Boyle, final cause could act in a realm parallel to the efficient mechanisms of his clockwork universe. Efficient causes pushed all the springs and cranked all the pulleys, but final causes expressed the purposes that God had in mind when he ordained the efficient mechanisms of his clockwork universe. God need not show his hand by miraculous intervention into the realm of efficient causation; Boyle's God is manifest in the final causes of phenomena constructed by ordinary efficient causes under nature's invariant laws.

But which phenomena of nature are best suited to the discovery and elucidation of their final causes? The logic of Boyle's presentation leads us squarely to organisms and their good design – in short, to adaptation as the quintessential natural phenomenon for displaying God's existence and attributes. In the *Disquisition* Boyle begins by citing the two major philosophical opponents to final causation: the Epicurean belief that a random universe can manifest no purpose, and the Cartesian claim that God's ends are too ineffable for human comprehension:

> Two of the chief sects of the modern philosophers, both of them, though upon differing grounds, deny that the naturalist ought at all to trouble or busy himself about final causes. For Epicurus, and most of his followers . . . banish the consideration of the ends of things; because the world being, according to them, made by chance, no ends of any thing can be supposed to have been intended. And on the contrary, Monsieur Des Cartes, and most of his followers, suppose all the ends of God in things corporeal to be so sublime, that 'twere presumption in man to think his reason can extend to discover them. So that, according to these opposite sects, 'tis either impertinent for us to seek after final causes, or presumptuous to think we may find them.
>
> (*Ibid.*, Preface)

Boyle then structures his search as a 'Goldilocks' problem, an attempt to find the 'just right' phenomenon between two extremes. He proposes three categories of natural objects that might manifest final causes: inanimate bodies of the cosmos, inanimate objects on earth and organic bodies on earth.

As the 'too big' category that will not illuminate final cause, Boyle identifies the immense cosmic bodies of the universe. Suns and planets must have divine purposes, but here Descartes may well be right, for we tiny inhabitants of one little planet will not be able to read God's purposes at so grand a scale. Cosmic bodies certainly show God's glory, but not his beneficence and care for us. 'The Cartesian way of considering the world, is very proper indeed to show the greatness of God's power, but not, like the way I plead for, to manifest that of his wisdom and benevolence' (*ibid.*, p. 37).

Into the 'too small' category of phenomena beneath God's adequate glory, Boyle places the inorganic phenomena of our scale on earth. For these bodies are too simple, and could well be formed either by chance as the Epicureans say, or by simple assembly following invariant laws of nature (God made the laws, of course, but final causes should illustrate his glory directly, not by one removal):

> As for other inanimate bodies, as stones, metals, *etc.*, whose matter seems not organized, tho' there be no absurdity to think, that they also were made for distinct particular purposes . . . yet most of them are of such easy and unelaborate contextures, that it seems not absurd to think, that various occursions and jostlings of the parts of the universal matter may at one time or another have produced them.
>
> (*Ibid.*, p. 44)

Boyle then nominates animals and plants as the 'just right' category for displaying the final causes that will illustrate God's existence and attributes to us – in other words, as the favoured objects of natural theology. Boyle cites three major reasons for this preference. First of all, organisms are so complex that we cannot attribute their forms and behaviours either to chance or to simple construction by nature's laws without overt and particular purpose:

> There are some effects, that are so easy, and so ready, to be produced that they do not infer any knowledge or intention in their causes; but there are others, that require such a number and concourse of conspiring causes, and such a continued series of motions or operations, that 'tis utterly improbable, they should be produced without the superintendency of a rational agent, wise and powerful . . . I never saw any inanimate production of nature, or, as they speak, of chance, whose contrivance was comparable to that of the meanest limb of the dispicablest animal: and there is incomparably more art expressed in the structure of a dog's foot, than in that of the famous clock at Strasburg.
>
> (*Ibid.*, pp. 45–7)

Second, organisms exist at our scale, and operate much as we do – so we can readily grasp the final causes of their design (as we may not for immense and fiery bodies so distant from the earth as other suns). Boyle focuses his attention on the classic case of adaptation: the design and function of eyes:

> The great author of things . . . has furnished various species of animals with

> organs of sight that are very differingly framed and placed . . . This diversity
> nobly manifests his great providence, and (if I may so call it) forecast, that has
> admirably suited the eyes of the differing kinds of animals, both to the rest of
> their bodies and . . . to those parts of the great theater of the world on which he
> designs that they shall live and act.
>
> (*Ibid.*, pp. 58–9)

Third, as also illustrated in the quotation above, we can readily grasp the *utility* of organic form and function – and final causes are most clearly manifest in function, or adaptation. Boyle follows the strategy, classic ever since among adaptationists of either creationist or evolutionary persuasion, of discussing apparent exceptions – features that seem degenerate or devoid of function – and then showing that these parts, as well, are optimally suited for a creature's particular mode of life. Boyle discusses the rudimentary eyes of moles:

> The eyes which nature hath given them, are so little, in proportion to their
> bodies, that 'tis commonly believed, and even by some learned men maintained,
> they have none at all. But though by anatomy, I, as well as some others that
> have tried, have found the contrary; yet their eyes are very differing from those
> of other four-footed beasts. Which is not to be wondered at; considering, that
> the design of nature was, that moles should live under ground, where a sight
> was needless and useless; and where greater eyes would be more exposed to
> danger: and their sight, as dim as 'tis, is sufficient to make them perceive that
> they are no longer under ground, . . . which seems to be the most necessary use
> they have of light and eyes.
>
> (*Ibid.*, p. 60)

Substitute natural selection for God's foresight, and a Darwinian adaptationist would not formulate the argument about function much differently.

The continuity in adaptation between Boyle's natural theology and Darwinism

With many exceptions to be sure, organisms do tend to be well designed, and though Darwin inverted Boyle's explanation, the phenomenology endures. Consequently, the power of the adaptationist programme remains unchanged from Boyle to modern Darwinism – as seen most clearly in the eminently operational and highly fruitful strategy of assuming good operation 'for' some function when trying to analyse an enigmatic structure. Boyle's brilliant passage on how William Harvey used the structure of venal valves to infer circulation of the blood beautifully illustrates Louis Pasteur's famous quip that 'fortune favours the prepared mind', and illustrates how assumptions of good design can work as admirable preparation. Substitute 'natural selection' for 'so Provident a Cause', and the heuristics of modern adaptationism shine forth just as well in this passage. Modern defenders of adaptationism (see, for example, Ernst Mayr's article for *American Naturalist*, 1983) locate a primacy rationale in the same utilitarian argument:

> I remember that when I asked our famous Harvey, in the only discourse I had with him, (which was but a while before he died) what were the things that induced him to think of a circulation of the blood? He answered me, that when he took notice that the valves in the veins of so many several parts of the body were so placed that they gave free passage to the blood towards the heart, but opposed the passage of the venal blood the contrary way: he was invited to imagine that so provident a cause as nature had not so placed so many valves without design: and no design seemed more probable than that, since the blood could not well, because of the interposing valves, be sent by the veins to the limbs; it should be sent through the arteries and return through the veins, whose valves did not oppose its course that way.
>
> (*Disquisition*, pp. 157–8)

But just as the method works so powerfully when the mechanics of good design can be thus exposed, notable foibles and weaknesses appear (for Boyle in 1688, as for adaptationists today) when the presumption of optimal function becomes a dogma asserted *a priori*, and a claim irrefutable in principle. The two most frequent critiques of uncritical and overextended adaptationist arguments today may also be applied to several of Boyle's examples, thus establishing a continuous pedigree across this greatest of intellectual divides, from creationist to evolutionary biology.

'Just so stories', with cleverness, rather than empirical support, as a primary criterion

Harvey's functional argument triumphed both because he could obtain confirming evidence (and could use his fruitful hypothesis to direct the search), and because he was right in his claim. In other and opposite uses – tactics that stymie investigation rather than inspire testing – adaptationists, all too frequently, simply propose a cunning explanation, and then consider their work accomplished by cleverness of argument, rather than empirical validation of claims. Many critics have referred to these proposals as 'just so stories', invoking

Rudyard Kipling's purposely fanciful explanations of how elephants got long trunks, and rhinoceroses wrinkled skin.

Consider a sequence of Boyle's claims in this mode, moving from the potentially valid to the demonstrably false. All fall into the pre-eminent category of adaptationist puzzles – structures that seem ill designed or devoid of function, but that none the less require adaptive explanation if the paradigm be as universally valid as supporters desire:

(1) *For transient non-utility*: How can a functional argument be constructed for embryonic structures that appear only transiently, and have no known intrauterine function. Boyle makes the clever argument, probably correct in this case, that such features are structural prerequisites (he calls them 'scaffolds') for the functioning organs that will follow:

> 'These temporary parts were framed by a forecasting, as well as a designing, agent, who intended they should serve for such a turn, and then be laid aside; it being utterly improbable, that an undesigning agent should so appositely and exquisitely frame scaffolds for the future buildings, if he did not beforehand destinate both the one and the other, to concur to the same ultimate effect.'
>
> (*Ibid.*, p. 167)

In this case, we must also recognize that, to Boyle and his contemporaries, a documentation of adaptive value for some future state – scaffolding for the later building – seemed a particularly powerful argument for a conscious designer: for how else could a structure arise only to bolster later utility? Our age has found a different solution in the concept of programmed instructions and their evolution. But Boyle's generation scarcely possessed even a metaphor for such a notion, except, perhaps, in toys like the music box. The introduction of the Jacquard pattern weaving loom in the eighteenth century, and of computer technology in our own times, has made this concept among the most familiar in modern life. We can all grasp the workings of DNA, and no scientist would now claim that organic construction for future ontogenetic utility implies anything about conscious design.

(2) *For apparent redundancy*: Moving to the less plausible though not ridiculous, and still entirely in the speculative mode, Boyle completely ignores the obvious structuralist alternatives and argues for a purely functional explanation of bilateral symmetry as insurance against loss:

> 'There seems to have been care taken, that the body of an animal should be furnished, not only with all things that are ordinarily necessary and convenient, but with some superabundant provision for causalities. Thus, tho' a man may live very well, and propagate his kind, (as many do), tho' he have but one eye; yet nature is wont to furnish men with two eyes, that, if one be

destroyed or diseased, the other may suffice for vision ... In short, nature has furnished men with double parts of the same kind, where that duplicity may be highly useful.'

(Ibid., p. 143)

(3) *For apparently overt harm:* Boyle argues, exposing another and still pervasive aspect of cultural bias in the ranking of our genders, that the anatomical weakness of a woman may not be 'good' for her individual life, but benefits the species as a whole in aiding procreation:

> 'Those of the female sex are not so happily framed, in order to their own welfare, as those of the masculine: since the womb, and other things peculiar to women, which are not necessary to the good of individual persons, but to the propagation of their species, subject that tender sex to a whole set of diseases, belonging to them either peculiarly, as they are women, or as they are with child, or brought to bed; from all which men are exempt ... Men [*now meaning all people*] may sometimes mistake, when they peremptorily conclude, that this or that part of an animal must, or cannot, have been framed for such an use, without considering the cosmical, and therefore primary and over-ruling ends, that may have been designed by nature in the construction of the whole animal.'

(Ibid., p. 220)

(4) *Guesses that turn out to be just plain wrong:* This general tactic suffers most when later discovery exposes earlier proposals as no more than fatuous guesswork. The last quotation provides one example, for we now know that women exceed men in life expectancy – so Boyle's 'sure' knowledge of divine intent suffers inversion. As another example, cleaner for exposing foibles of the method, though clearly of much less social import – Boyle apparently thought, quite wrongly, that human teeth continued to grow throughout life, and he proposed divine utility for this non-existent phenomenon:

> ' 'Tis considerable, that whereas, when man is come to full stature, all the other bones of the body cease to grow, the teeth continue to grow in length during a man's whole life ... Of the difference in point of growth betwixt the teeth and other bones, what reason can be so probably given, as, that 'tis designed to repair the waste that is daily made of the substance of the teeth, by the frequent attritions that are made, between the upper and lower tier in mastication.'

(Ibid., p. 182)

Switching only within the paradigm upon falsification

Critics would not object so strongly to adaptationist arguments as invariable first approaches if falsification of a particular claim could lead to tests of truly different alternatives outside the adaptationist programme. But the committed

functionalist does not work in so open a manner, and disproof of one adaptationist hypothesis leads only to lateral feinting towards a different story, still
invariably in the functional mode. Thus, the paradigm cannot be refuted from
within.

We already encountered one example of this 'unbeatable' strategy above,
when Boyle, puzzled by the supposed bodily weakness of women, and meeting
difficulty in contriving a functional argument based on advantages for individual women, simply switched levels within the paradigm by arguing that any
detriments to women as individuals must be outweighed by benefits accruing
to the entire species.

Boyle uses this tactic throughout his exposition. After praising the ubiquity
of good biomechanical design, for example, Boyle falters when he cannot find
a functional explanation for vibrant animal colours (ironic, since we have now
taken most cases into the paradigm under the guise of sexual selection). But he
does not abandon functional explanation; instead, he switches, arguing that colours must be 'for beauty', rather than biomechanical utility:

> It may help us if it be considered, that, since God is both a most free and a most
> wise agent, it need not seem strange that he should adorn some animals, with
> parts or qualities that are not necessary to their welfare, but seem designed for
> their beauty: such as are the disposition of the camelion to change colours; and
> the lovely greens, blews [sic], yellows and other vivid colours that adorn some
> sorts of pigeons, and of parrots . . . and especially those admirable little winged
> creatures humming birds.
>
> (*Ibid.*, p. 205)

Extending the same argument to the even more troubling fact of riotous
diversity upon common designs (why do so many kinds of eyes exist, when the
basic structure works so well, and when most variants have no obvious link to
particular modes of life), Boyle floats the peculiar proposal – and do I, for once,
detect just a trace of timorousness on his part for the 'reach' of his special plead-
ing – that God uses this variety to instruct us about the range of His wisdom. (I
find this argument all the more forced because ordered diversity not linked to
particular life styles provides the strongest phenomenology for the alternative
structuralist approach based on 'laws of form' and the regularities of
transformation):

> If that be admitted . . . as very likely, that God designed, by the great variety of
> his works, to display to their intelligent considerers, the fecundity (if I may so

> speak) of his wisdom; one may readily conceive that a great part of the variety observable in the analogous parts of animals, as their eyes, their mouths, *etc.* may be very conducive to so reaching and comprehensive a design; to which the beauty of some creatures and parts, as well as their more necessary or convenient structure, may be subservient; especially if the innocent delight of man be also intended, as it may seem to be in the curious colours and shapes of divers flowers, and in the melodious musick of singing birds, and in the vivid and curiously variegated colours of the feathers of several winged animals, particularly those that make up the peacock's train.
>
> (*Ibid.*, pp. 208–9)

To end this section with Boyle's most explicit affirmation of exclusive adaptationism as a methodology, he defends both tactics critiqued above – reliance on conjectural 'just so stories', and switching only within the paradigm upon refutation of particulars – by stating that we poor mortals cannot grasp the full range of God's intent, and that our failures to discern function point more surely to our ignorance than to the propriety of an alternative explanation:

> Men may easily be too rash, if they think a part bunglingly framed, upon supposition that, by the anatomical inspection of it, they know all the uses that the skill of the divine opificer could design for it.
>
> (*Ibid.*, p. 203)

> Tho' we may safely conclude that God acts wisely, when he does something that has an admirable tendency to those ends we justly suppose him to have designed; yet we cannot safely conclude in a negative way, that this or that is unwise, because we cannot discern in it such a tendency. For so wise an agent may have other designs than we know of, and further aims than we can discern or perhaps suspect . . . [aims that are] far above the reach of our conjectures, and without the knowledge of which we but rashly censure the wisdom of his proceedings.
>
> (*Ibid.*, pp. 209–10)

A truly unbeatable argument but, for that reason (or so we would say today), not very useful in science!

The radical difference between created and evolutionary (particularly Darwinian) adaptationism

Darwin frequently stated that he had tried to advance two quite separate innovations in proposing his theory of 'descent with modification': first, simply to convince people of evolution's factuality (implying the genealogical basis of organic relationships and the transmutational interpretation of life's history); and, second, to propose a theory (natural selection) for the causes of this factuality. He then added that the first aim – establishment of factuality rather than confirmation of mechanism – must be viewed as far more important because the revolutionary consequences of such an admission ran so deep and so counter to Western traditions. We may use this famous, and wonderfully perceptive, self-assessment to identify what changes so radically amidst the continuity of adaptationist argument in anglophonic natural history.

Beginning with the second aim of establishing theory, many historians have noted that the most revolutionary feature of Darwin's mechanism lies in its almost brutal inversion of natural theology. For Boyle, adaptive design represents the direct handiwork of a caring God; for Darwin, the same phenomenology emerges as a side-consequence of a causal principle that could carry only opposite moral messages, if morality could be read in nature (as, according to Darwin, it most definitely cannot) – namely, a struggle among individual organisms for personal reproductive success.

This crucial inversion imposes a difference in the manner of adaptationist argument pursued by Boyle and later Darwinians – for Boyle may locate adaptation at any level of biological organization (as all can display God's intent), while Darwin must ascribe benefits only to organisms in their reproductive competition, and must therefore deny such 'cosy' concepts as the 'good of the species'. Boyle, as we have seen in his argument about the supposed weakness of women, happily switched to adaptive benefits for entire species when he could not identify advantages for individuals. In an even more telling example, Boyle recognizes the placenta as excellent design, but can only attribute the structure to divine beneficence for our entire species, since the health and strength of individual women are not enhanced thereby (and Darwin's key notion of individual *reproductive* success as an organic *summum bonum* is understandably absent from Boyle's mental map of nature):

> Those temporary parts appear to have been designed by nature, not so much for the personal preservation of the female as for the propagation of the species: which destination . . . appears to have been preordained by the author of mankind for the continuation of it [that is, the species].
>
> (*Ibid.*, p. 152)

However striking this difference imposed by theory, far more portentous changes were enjoined (as Darwin had correctly noted) in accepting Darwin's first aim of establishing the basic factuality of evolution. After all, Darwin's own insistence upon struggle among individuals as the only proper level for arguments based on natural selection – however brilliantly affirmed in the twentieth-century history of evolutionary thought – was quite idiosyncratic to Darwin, and so radical that most of his contemporaries, including his strongest supporters, never understood the depth of meaning involved in this restriction in levels. Darwin's colleague and co-proponent of the theory of natural selection Alfred R. Wallace, for example, was quite content to make conjectural arguments about natural selection at all levels, including frequent claims about the 'good' of species. Thus, Wallace's evolutionary arguments might not have differed from Boyle's defences for design on this score.

But the simple admission that organisms have histories of genealogical connection imposes such a different geometry upon the structure of life that all arguments in natural history must alter. (This grand change has minimal effect upon claims about good design for *particular* creatures in one environment at *one time* – hence the maximal continuity between Boyle and Darwinism on questions of adaptation.)

At the broadest scale, genealogical thinking opens the insight that many anatomical features must be simple signs of ancestry and need not be elaborately explained, part by individual part, as adaptations expressly crafted for current function. As Boyle struggles to explain why bats, uniquely among 'birds' (used as a functional term for flying vertebrates, not as a genealogical designation), have so many characters otherwise found only in furry creatures on the ground, one almost wants to kick him and say: 'don't you see; it's so simple; bats are mammals by ancestry' – but then, of course, Boyle could not see this solution, and the role of world views in both constraint and facilitation lies beautifully exposed in the example.

> Though bats be looked upon as a contemptible sort of creatures [*sic*], yet I think they may afford us no contemptible argument to our present purpose. For in the heteroclite [*an archaic word meaning anomalous*] animal, you may discern the fecundity of the divine artificer's skill, which has in this formed an animal that flies like birds, and yet is not only unfurnished with feathers but is of a fabric quite differing from that of other birds. And in this little animal we may

> also observe . . . the regard, which the divine artist appears to have to the symmetry of parts, in his animated works, and to their fitness for the places they are to live in or frequent. For the bat, being to act sometimes like a bird, that flies freely to and fro in the air, and on some occasions like a terrestrial animal, such as is that little quadruped a mouse; ought to be furnished with parts suitable to such different destination.
>
> (*Ibid.*, pp. 193–4)

In a more subtle difference on the same point, construction by sequential history, rather than creation in full perfection, immediately resolves a problem that Boyle found quite puzzling (though not insurmountable by his cleverness): how can we interpret a function that has current utility to an organism, but that must be explained as a secondary consequence of a different primary or original usage – the thin coin minted for currency, but employable as a screwdriver in moments of need. Boyle, lacking a concept of historical change, has to argue that his excellent divinity also foresaw *all* the secondary utilities when he created the feature for a primary role:

> I have seen, and been master of a telescope, made in the form of a walking-staff, so that it was fitted to serve for several purposes; whereof tho' one was very different from the other, yet all of them were in the idea of the artificer, and intended by him.
>
> (*Ibid.*, p. 99)

Think of the intellectual liberation supplied by the stunningly simple alternative that adaptations evolved for one function may be fortuitously fitted to work in other ways as well – so that feathers arising as thermoregulatory devices may then be co-opted for flight. Liberating, but immensely threatening (and therefore invisible for Boyle) to a belief in a young and static world, replete with final causes displaying the existence and benevolent intent of an omnipotent deity.

The importance of conceptualizing alternatives

From my extensive quotation of Boyle, and with the benefit of our Darwinian insight and hindsight, we can easily perceive the confines of his seventeenth-century conceptual prison. Boyle's natural world contains no historical dimension, and he must therefore view every mammalian feature of a bat as expressly created for a current function, and not as a mark of ancestry. His view of nature proclaims ubiquitous purpose (illustrating God's benevolently creative order), and he must therefore be stymied (or driven to forced conjecture) by a range of phenomena differently and properly rendered in systems ordered by genealogy (features co-opted for secondary functions, vestigial organs, utilities based on criteria – particularly the Darwinian *summum bonum* of reproductive success – unimportant to Boyle's God).

But we would be traitors to the scholarly imperative of search for under-standing if we read Boyle's differences from our view of life as an excuse to bewail his opacity, or as a way to exalt our own sophisticated times against his 'bad old days' (I doubt whether any of us could hold a candle – to cite a meta-phor of past technologies – to Boyle's raw intellectual power). Rather, we should read the obvious lesson in intellectual constancy. If such a brilliant man dwelled in a conceptual prison so patent to us today, how are we unwittingly incarcer-ated in systems of belief that will seem just as ludicrous and easily discarded to our descendants?

By tracing the continuity in anglophonic evolutionary thought between Boyle and modern Darwinism on the crucial subject of adaptation, or func-tional explanation in general, I wish to suggest that we might profitably exam-ine this ancient preference – for our propensities may be recording equal parts of nature's factuality and our own continuing conceptual prisons.

Alternative approaches to evolutionary theory do exist, often with long pedi-grees in largely continental traditions; evolution is not co-extensive with adap-tationist preferences in explanation. Adaptation will always be a vital subject in evolutionary thought, for organisms do tend to be well designed, and natural selection is a proven and potent force. But adaptation need not be the funda-mental result of evolution's causal workings, the pre-eminent and controlling phenomenon of life's transmutational history. Perhaps the continental perspec-tive is more correct, and most adaptations rank as subsequent, particularistic modifications of underlying structures and as products of their transform-ational rules and regularities.

This forum is not the place for an extensive compendium, or a long defence, of these alternatives. (As an agnostic on this issue, I would not even be comfort-able in presenting such a defence, nor can we fairly depict the issue in such a dichotomous manner at all.) But I do think that a variety of structuralist approaches are now in the ascendancy, thus giving new life to an old division that goes back to the pre-evolutionary version of Geoffroy Saint-Hilaire touting the power of archetypes and laws of form against the non-evolutionary adaptationism of Georges Cuvier in their famous debates of 1830 at the Académie des Sciences in Paris. D'Arcy Thompson kept the structuralist vision alive, with an explicitly anti-Darwinian evolutionary version, in the finest work of prose in English natural history –*On Growth and Form*. This decade, Stuart Kauffman and Brian Goodwin have both written powerful and provocative, if flawed, modern versions that explore sources of biological order arising from structural rules rather than functional selection. (Kauffman, in particular, has underlined the potentially non-oppositional status of structuralism to Darwinian functionalism, pointing out that his laws of form provide order 'for free' to a selective system that can then modify and add further regularity.) Most spectacularly, our stunning successes in beginning to unlock the genetics of development have proven a depth of structural constraint by homology across the most disparate of complex phyla, particularly arthropods and chordates, in body axes, substrates for the formation of eyes and segmental differentiation. Most amazingly (I had long rooted for such a result after writing *Ontogeny and Phylogeny* in 1977, but never dared really to hope for a positive outcome), it begins to apear that Geoffroy Saint-Hilaire was correct in his homology of the vertebrate body plan to an inverted arthropod design, for homologous determinants of dorso-ventral patterning are indeed reversed in the two phyla. Details can be found in the articles by Y. Sasai *et al.* and S. A. Holley *et al.* in the list of Further reading at the end of the chapter.

In this light, one might ask why an evolutionist should worry about the dogmatism of overly strict adaptationism; will this viewpoint not be swept aside by the successes of modern structuralist thinking, thereby leading all evolutionists to a proper pluralism? Apparently not, and ironically, for strict Darwinian adaptationism – now given the quite appropriate name of 'ultra-Darwinism' by B. Goodwin and N. Eldredge – remains strong in anglophonic evolutionary circles (whether by sheer vestigial weight of an adaptationist tradition dating back to the seventeenth century, or by the attraction so many of us seem to feel for simplistically comprehensive world views, I do not know. I do not think that either power of evidence or strength of argument can be supporting such an exaggerated and one-dimensional theory).

I am not much concerned about the fallacies of ultra-Darwinism within evolutionary biology, for most professionals understand the limitations of such a view only too well – and the current leading exponent, Richard Dawkins, seems to maintain a strict attachment to the creed that can only be called theological. I worry more when practitioners of other disciplines dip into evolutionary biology, see only this traditional (if superannuated) viewpoint, fall in love with its beguiling simplicity and then make the great error of thinking that they have accurately translated another field into their own.

Thus, for example, the philosopher Daniel Dennett extols the ultra-Darwinian straight and narrow, while a caricature of the true richness of Darwinian functionalism passes as a paradigm for a 'new' discipline self-consciously touting itself as 'evolutionary psychology' (see D. M. Buss, 1995, for a technical account of this field, and R. Wright for a sycophantic 'pop' version). Evolutionary psychologists view themselves as 'sophisticated' about adaptation because they do not argue, as some even more naïve sociobiologists did in the last round of discussion, that all behavioural universals must be adaptively maintained. These new apostles of ultra-Darwinism hold that many universals have become tragically non-adaptive in modern society, but must have been adaptive at their origin on the African savannahs (or wherever) – for natural selection is the cause of evolution and natural selection builds adaptation. Thus, the evolutionary psychologists remain thoroughly ultra-Darwinian in positing an adaptive origin for all human universals – while true alternatives require recognition of the richness of non-adaptative means whereby such universal traits may arise – see S. J. Gould and R. C. Lewontin on the principle of spandrels, and other potent non-adaptive mechanisms that must be largely responsible for the uniquely human utilities of our mental functioning.

Darwin's own position in this continuing debate over so many centuries remains powerfully relevant and of far more than mere historical interest. As a subtle thinker, who knew that the richness of natural history could not yield to one-dimensional explanation, but who cherished the power of his own intel-lectual issue of natural selection, Darwin both overstressed the anglophonic preference for adaptationism that defined his patrimony, and at the same time, warned against too exclusive a reliance on this single mode. In fact, nothing could call forth more annoyance from this remarkably genial man than the dis-tortion of his theory into a cardboard version that equates natural selection with the exclusivity and omnipotence of Boyle's deity (and, on this ground, I am confident that Darwin would have eschewed ultra-Darwinism). For example, he wrote in near despair for the last edition of the *Origin of Species* (1872, p. 395):

> As my conclusions have lately been much misrepresented, and it has been
> stated that I attribute the modification of species exclusively to natural
> selection, I may be permitted to remark that in the first edition of this work,

and subsequently, I placed in a most conspicuous position – namely at the close of the Introduction – the following words: 'I am convinced that natural selection has been the main, but not the exclusive means of modification.' This has been of no avail. Great is the power of steady misinterpretation.

And yet, Darwin was not a pluralist without preference. His basic world view elevated the functional mode above all others by defining adaptation as the central problem of evolution (see the set of quotations on pp. 4–7). In so doing, he expressed his fealty to a national tradition extending right back to Boyle and his compatriots at the foundation of modern science.

When I last spoke at Darwin College, for the grand celebration held to commemorate the centenary of Darwin's death, I ended my presentation with an incisive line from William Bateson, a great non-Darwinian evolutionist who none the less caught the essence of Darwin's paramountcy among English scientists. As I have written an article about continuity across centuries and through the greatest of all intellectual transformations in the history of biology, may I end with a small personal continuity in citing Bateson once again, and in the same position – for his words ring with the explanatory pluralism that we will have to champion if we wish to fathom the complexities of evolution:

> We shall honour most in him not the rounded merit of finite accomplishment, but the creative power by which he inaugurated a line of discovery endless in variety and extension.
>
> (Bateson, 1909)

FURTHER READING

Bateson, W. 'Heredity and variation in modern lights'. In *Darwin and Modern Science*, A. C. Seward (ed.), pp. 85–101, Cambridge: Cambridge University Press, 1909.

Boyle, R. *A Disquisition About the Final Causes of Natural Things, Wherein it is Inquir'd Whether, and (If at All) With What Cautions, a Naturalist Should Admit Them*, London: John Taylor, 1688.

Buss, D. M. 'Evolutionary psychology: a new paradigm for psychological science', *Psychological Inquiry* 6 (1995), 1–30.

Darwin, C. R. *The Origin of Species*, London: John Murray, 1859.

Darwin, C. R. *The Origin of Species*, 6th edition, London: John Murray, 1872.

Dawkins, R. *A River Out of Eden*, London: Weidenfeld & Nicolson, 1995.

Dennett, D. *Darwin's Dangerous Idea*, New York: Simon & Schuster, 1995.

Eldredge, N. *Reinventing Darwin*, New York: J. Wiley, 1995.

Goodwin, B. *How the Leopard Changes Its Spots*, London: Weidenfeld & Nicolson, 1994.

Gould, S. J. *Ontogeny and Phylogeny*, Cambridge, MA: Belknap Press of Harvard University Press, 1977.

Gould, S. J. 'Exaptation: A crucial tool for an evolutionary psychology', *Journal of Social Issues* **47** (1991), no. 3, 43–65.

Gould, S. J. and Lewontin, R. C. 'The spandrels of San Marco and the Panglossian paradigm: A critique of the adaptionist programme', *Proceedings of the Royal Society of London B* **205** (1979), 581–98.

Holley, S. A., Jackson, P. D., Sasai, Y., Lu, B., De Robertis, E. M., Hoffman, F. M. and Ferguson, E. L. 'A conserved system for dorsal–ventral patterning in insects and vertebrates involving *sog* and *chordin*', *Nature* **376** (1995), 249–53.

Kauffman, S. *The Origins of Order: Self-Organization and Selection in Evolution*, New York: Oxford University Press, 1993.

Mayr, E. 'How to carry out the adaptationist program?', *The American Naturalist* **121** (1983), no. 3, 324–34.

Paley, W. *Natural Theology*, London: R. Faulder, 1802.

Sasai, Y., Lu, B., Steinbeisser, H., Geissert, D., Gout, L. K. and De Robertis, E. M. '*Xenopus chordin*: a novel dorsalizing factor activated by organiser-specific homeobox genes', *Cell* **79** (1994), 779–90.

Thompson, D'Arcy W. *On Growth and Form*, London: MacMillan, 1917.

Thompson, D'Arcy W. *On Growth and Form*, 2nd edition, London: MacMillan, 1942.

Wright, R. *The Moral Animal: The New Science of Evolutionary Psychology*, New York: Pantheon, 1994.

2 The Evolution of Cellular Development

LEWIS WOLPERT

The evolution of the cell is nature's greatest evolutionary triumph. That may sound rather presumptious when one compares the apparently humble cell with the complexity of organisms such as human beings, with their extraordinary brains. But in evolutionary terms, it was only once the cell had evolved that multicellular organisms became possible. Moreover, I suggest that given the eukaryotic cell – which has a nucleus, contains organelles such as mitochondria, is capable of movement and is itself evolved from the simpler bacteria – the evolution of complex structures, even the brain, was by comparison relatively simple.

Evolution as the modification of development

Development is central to the evolution of multicellular organisms: evolution proceeds by the modification of the embryo's developmental programme to produce differences in the adult. This modification is due to changes in the genes controlling development; they act by controlling cellular behaviour during development. It is by this process that, to use the French molecular geneticist François Jacob's phrase, evolution can tinker with embryos, using its bits and pieces to make new structures. This can be seen very clearly when one looks at the early embryos of vertebrates: they all look remarkably similar at one stage (known as the phylotypic stage) and then diverge (Figure 1). Evolution has tinkered with the basic body plan.

A clear example is the limb. While the basic form of the limb has been retained in many land vertebrates, its development has been 'tinkered with' to provide the bird's and bat's wings, the horse's leg and our own manipulative hand. Moreover, the limb itself evolved from the fins of fish. Another dramatic example is the middle bone of the middle ear in mammals. This bone was originally, in our reptilian ancestors, part of the articulation of the jaw. And

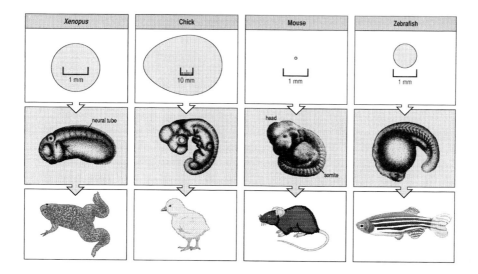

FIGURE 1 Vertebrate embryos converge on a common phylotypic stage and then diverge. Top, fertilized egg; middle, phylotypic stage; bottom, adult.

another example is provided by the genitourinary system in which the primitive kidney has become modified to become part of the reproductive system, and a separate kidney has evolved.

Modification of growth has also been a major feature of evolution. This can be seen in the evolution of the horse's leg, where, because lateral digits grow more slowly than the central ones, as the horse increases in size the lateral digits become much smaller relative to the central ones, making the central digit the main load carrying one. Again, the changes in the shape of vertebrate skulls is largely due to growth (Figure 2). As D'Arcy Thompson pointed out in *On Growth and Form*, some eighty years ago, differential growth can account for the overall differences in the shapes of skulls.

Thus, evolution can be seen to be the modification of development. We need then to understand development and how development itself evolved.

Development

The link between cells and multicellular organisms is, with rare exceptions, embryonic development from a single cell, the egg. (An exception is the cellular slime mould, which develops by the aggregation of a large number of amoebae.) All the animals and plants we see around us have developed from this single

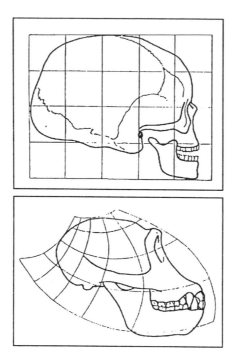

FIGURE 2 The shapes of human (top) and ape (bottom) skulls largely reflect differences in growth.

cell, the fertilized egg. Its development is characterized by a series of cell divisions which result in a mass of smaller cells – usually of the order of hundreds or thousands. This early mass of cells then gives rise to a variety of different cell types such as muscle and cartilage, a process known as differentiation. These cells give rise to well-defined patterns, as for example in limbs, and this is known as pattern formation. The differences between vertebrates, between birds, humans, hippopotamuses and chimpanzees lie primarily in pattern formation – how the cells are spatially organized rather than due to significant differences in the cell types that make them up. We do not, for example, have any particular cell type that chimpanzees lack. A third process involves change in form in which, for example, sheets of cells change their shape. In gastrulation cells that will ultimately form the gut and skeleton and which are on the outer surface of the embryo move inside the embryo. All patterning of the embryo occurs when the embryo is quite small – a human limb is patterned when it is much less than a centimetre long – and it is growth at later stages that produces large structures. Not only is all patterning carried out with relatively small numbers of cells, but the interactions between cells are quite local; once the basic body plan is laid down, the development of particular organs such as eyes and limbs is relatively autonomous.

It is the behaviour of the individual cells that determines the outcome of embryonic development. DNA, the genetic material, controls development and it does so by controlling cell behaviour. DNA is, ironically, rather passive. What DNA contains is the code for making proteins, and it is proteins that determine how a cell behaves. Enzymes that catalyse the key chemical reactions in the cell are proteins, and so are many of the structural components, like those of the cell cytoskeleton and membrane. A cell is characterized by the proteins it contains, especially the luxury proteins – keratin in skin, and haemoglobin in red blood cells – as distinct from the housekeeping ones that most cells share and which underpin basic functions. A particularly important class of proteins comprise transcription factors, which control which genes are switched on and which are switched off and therefore eventually which proteins a cell makes. A cell's state or character is thus determined by which genes are on or off and hence which proteins are present. The DNA of the embryo does not provide a blueprint for the adult organism, it is much better to think of it as containing a set of instructions for making the organisms by controlling the sequence of protein synthesis. A good analogy is paper folding or origami. In development, the instructions are read out in terms of the synthesis of specific proteins as development proceeds.

I now describe some of the basic pattering mechanism involved. For, in order to understand the evolution of development, one needs to appreciate the major features of embryonic development.

One mechanism for making patterns is based on positional information (Figure 3). The basic idea of positional information is that the cells acquire positional identities with respect to certain boundaries, rather like specifying position with a system of co-ordinates. Having acquired a positional identity, or address, the cells can then interpret this by a particular behaviour which is also influenced by the cell's developmental history. This change in cellular behaviour may result in differentiation of a specific cell type or lead to change of form by, for example, altering cell shape or adhesiveness. Interpretation may also involve specifying a programme of growth. A key characteristic of such a

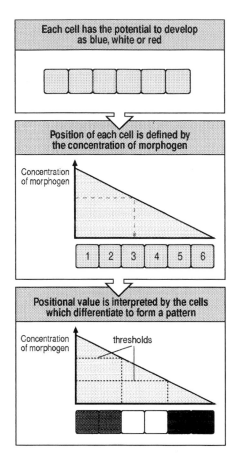

FIGURE 3 Positional information can be used to specify spatial patterns. Position could be specified by the amount of a 'morphogen' and the cells could read the local concentration.

mechanism for specifying pattern is that there is no simple relationship between positional specification and the final pattern, the two are dissociated. Indeed, one of the attractive features of positional information is that the same positional identities can be used to specify an enormous variety of patterns and there is now evidence to support this conjecture. The nature of each pattern will depend on how the cells interpret their positional identity.

Positional information could thus provide a unifying concept for understanding the development and regulation of a wide variety of patterns. The only cell-to-cell interactions that are, in principle, required are those necessary to specify position. Again, in principle, the same signals and positional identities could be used to generate quite different patterns, the differences reflecting both the cells' genetic constitution and developmental history.

Laying down the body plan

Early development involves laying down the body plan – that is, specifying the main axes (antero-posterior, and dorso-ventral) and specific regions (e.g. where the head and limbs will form). It provides a rough outline of the future organism and, in segmented animals, establishes the periodic pattern. Laying down the body plan in vertebrates results in the embryos passing through a common phylotypic stage (Figure 1). It is very difficult at this stage to distinguish between embryos of different vertebrates, and it is from this stage onwards that they diverge. There is also a divergence before the phylotypic stage as one traces development backwards to the egg. The early development of a mammal is quite unlike that of a frog but the embryos then converge to the phylotypic stage. It is an important evolutionary question as to why the phylotypic stage should have been so well conserved.

We need not concern ourselves with the details of very early development and can concentrate on the laying down of the body plan that leads to the phylotypic stage. Insects, too, have a phylotypic stage and we start with them since it is here that we have our best understanding of how genes control early development. The fruit fly *Drosophila* is, at present, the best model organism and understanding its early development has had profound implications for the study of other systems.

In *Drosophila* the axes (antero-posterior and dorso-ventral) are laid down in the egg. For example, there is a gene called *bicoid* which codes for a transcription factor and the messenger RNA transcribed from this gene is localized at the anterior end of the egg. After fertilization there is a gradient of bicoid protein along the antero-posterior axis (Figure 4). This gradient can be thought of as providing the cells of the embryo (though at this early stage there are no cell walls between nuclei) with positional information. That is, the cells, by reading the concentration of the bicoid protein, could, in principle, 'know' their distance from the anterior end. At particular threshold concentrations other genes are turned on and they, too, code for transcription factors which can activate and inhibit yet other genes. There is thus a cascade of gene activities that leads to

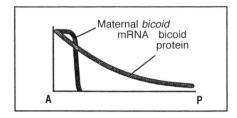

FIGURE 4 Positional information in the egg of the fruit fly *Drosophila* is initially specified by a concentration gradient in the protein bicoid, which is synthesized at the anterior (A) end of the egg and diffuses towards the posterior (P) end.

the regionalization of several protein transcription factors. These factors then activate other genes (the pair rule genes), each of which is active in seven 'stripes', just a few nuclei wide, which define fourteen parasegments, the fore-runners of the body segments of both the larva and the adult fly. Each stripe of activity of a pair rule gene is specified by the local concentration of transcription factors which act on the genes' control region.

At a slightly later stage, at what might be considered the phylotypic stage, each segment acquires a unique identity by virtue of the genes active in it. These special genes code for transcription factors which give the segments their positional identity and they belong to the so-called homeotic gene complex. The genes in this complex are linked together on the same chromosome and have a number of particular interesting features. First, their name, homeotic, derives from the fact that mutations in these genes can cause the alteration of one structure into another, which is known as homeosis. For example, there are mutations – changes in the genes – which result in a leg developing instead of an antenna, or which can alter the identity of one segment into that of a neighbouring segment so that a four-winged fly results. Secondly, the order of the genes along the chromosome is co-linear with the order in which they are expressed along the antero-posterior axis. Genes at one end of the complex are expressed at the anterior end of the embryo, while those at the other end are expressed in more posterior regions. Thirdly, all the genes contain a rather similar small region known as the 'homeobox', which codes for part of the protein that binds to DNA.

The presence of a homeobox is particularly significant since similar conserved regions have been found in almost all animals. In vertebrates there are four homeobox gene complexes and they, too, are expressed along the antero-posterior axis in the same order as they are arranged in the fly. It is generally thought that the genes of the homeobox complexes record positional identity along the main body axis. Once, it seems, evolution had found a satisfactory way of recording differences in position it used them again and again. All these gene complexes are derived from some primitive ancestor and in vertebrates the four complexes must have arisen by duplication of genes in an ancestor.

By what mechanism do the genes of the homeobox complex become specified in the correct position in vertebrates? Unfortunately, the answer is not clear, but it does involve signalling between the cells.

Limb and wings

Development of the vertebrate limb, once the body plan has been laid down, is important in its own right but is also a useful model system. The processes patterning the limb are very similar to those that lay down the body plan.

The chick limb starts off as a bud that grows out from the flank and the skeletal elements are laid down in sequence: humerus, radius and ulna, wrist and then digits. The digits are quite distinctive and are, by convention, named 2, 3 and 4, with 2 being the most anterior and 4 posterior. (In our hands the little finger is posterior.) We can focus on the patterning of the digits.

A region at the posterior margin of the limb bud, the polarizing region, seems to be the source of a signal. Grafting a polarizing region to the anterior margin of another limb results in a mirror-image limb (Figure 5a, b). So instead of a normal pattern of digits 2 3 4 the pattern is now 4 3 2 2 3 4. The polarizing region has signalled to the anterior region of the limb bud a new set of positional identities, possibly by a graded signal with digit 4 forming at a high threshold and digit 3 at a low one. Recent studies have shown that the gene *sonic hedgehog* is expressed in the polarizing region and its protein is a good candidate for a positional signal.

The same signal specifying pattern along the antero-posterior axis is present in both fore and hind limb buds and, indeed, all amniotes. Thus, a mouse polarizing region can specify mirror-imaging in the chick. Moreover, the *hedgehog* gene seems to be involved in positional signalling in other regions of the embryo, for example in specifying the main body axis and the pattern of neurones in the spinal cord. This suggests that the same positional signal can act at different times in development. It is the molecule of the moment.

35

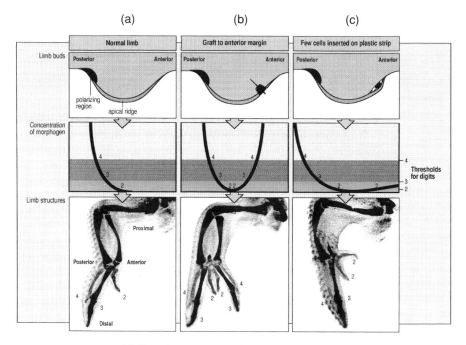

FIGURE 5 (a) The polarizing region of the chick limb can specify digit identity – the mechanism could be based on a diffusible morphogen produced by the polarizing region. (b) When an additional polarizing region is grafted to the anterior region a mirror-image limb results. (c) If fewer polarizing region cells are grafted, attenuation of the signal occurs and only a partial duplication results.

There are also striking similarities in the development of the wing of *Drosophila* and the vertebrate limb. Again, the *hedgehog* gene is involved and it is turned on in a narrow strip that runs along the midline of the developing wing. It appears to specify a diffusible 'morphogen' that patterns both anterior and posterior regions. So if *hedgehog* is expressed in an unusual position it is rather like a graft of a polarizing region and it specifies a new set of wing structures.

Conservation of developmental mechanisms

This very brief discussion of developmental processes has shown that, in systems as diverse as insects and vertebrates, common developmental mechanisms have been retained in evolution. The genes specifying cell state as well as the signalling molecules can be similar in different organisms. Amongst the best examples are homeobox gene complexes, and the families of signalling molecules such as hedgehog that are used again and again. A very recent and surprising example is that similar genes are involved in the development of both the insect and vertebrate eye.

This conservation among developmental mechanisms should not be used to conceal an apparent variety of developmental mechanisms, particularly at the earliest stages. Animals and plants adopt many different mechanisms for setting up the axes in the egg. The dorso-ventral axis of the amphibian embryo is specified by the site of sperm entry, whereas the mammalian egg is spherically symmetrical and its axes are laid down quite late. In the sea squirt the future muscle is specified largely by special cytoplasm rather than by cell interactions.

Nevertheless, in a sense, evolution has been quite conservative, even lazy, in its invention of developmental mechanisms. Once mechanisms that generated complex organisms were 'discovered' the mechanisms were used again and again. For example, patterning mechanisms based on positional information and generation of periodic structures generate an enormous, almost infinite, variety of patterns and structures, particularly when coupled with change in form and growth. There is nothing special about the development of the human brain, which, on the whole, makes use of just the same developmental mechanisms as those used by our primitive ancestors.

Origin of the embryo

The cell is a marvellously complex system. The interactions between its components are many orders of magnitude greater than between the cells of an embryo. In this sense the cell is much more complex than an embryo. Embryonic development is merely the co-ordination and modification of cellular activities.

The arrival of the cell some 3000 million years ago can thus be regarded as the 'Big Bang' of biological evolution, even though it took a very long time. The origin of embryonic development from cells can be regarded as the 'Little Bang', since the cell was already there. So the general question is, what was required, given the eukaryotic cell, for development to occur? How did the egg, patterning and change in form originate? Since embryonic development requires the formation of a multicellular organism from a single cell, the origin of the egg is a central problem.

The evolution of development is linked to the origin of multicellular forms – Metazoa. The earliest eukaryotic organisms are thought to have been present some 1400 million years ago, whereas the earliest evidence for metazoans is some 800 million years old. It may thus seem that the transition from single-celled eukaryotes to multicellularity was a difficult one, requiring hundreds of millions of years. However, the fossil record for those delicate primitive ancestors is undoubtedly fragmentary and incomplete. I suggest, given the eukaryotic cell with its ability to replicate and move, that all the basic elements required for development were already present, and the transition to multicellularity was relatively easy. Even so it took another few hundred million years until the Cambrian explosion, when fossils of recognizable animals can be found.

The basic processes in development (differentiation, spatial patterning, change in form and growth) were already present in the eukaryotic cell. The cell cycle, i.e. the sequence of events that results in cell growth and division, illustrates all these processes.

During the cell cycle there is a sequence of gene activity that can be regarded as a simple predecessor of a developmental programme. As the cell grows, genes are turned on and off at different stages in the cell cycle, and these stages can be thought of as being like differentiated states. There is also cell movement at cell division and also spatial patterning that ensures the chromosomes are distributed equally to the two daughter cells. There are, in addition, thresholds involved in the transition from one phase to the next, as control substances accumulate. The mechanism of cell division, mitosis, also provided the opportunity to make the daughter cells different by distributing cell components unequally at cell division. Indeed, this trait is present in yeasts, where the daughter cells may be different with respect to mating.

Here, I try to present a scenario – a 'just so story' – whereby the eukaryotic cell could have evolved multicellular embryonic development. A central requirement of this is that each stage must have a selective advantage and that there is continuity between stages. Big jumps – hopeful monsters – are not allowed. Even so, I recognize that my scenario is only slightly better, perhaps, than one of Rudyard Kipling's *Just So* stories, like 'How the leopard got its spots', or 'How the camel got its hump'.

There are two main theories as to the origin of the Metazoa and hence the origins of development. The first proposes that there was a coming together of two or more cells to form a colony while the second proposes growth of a cell

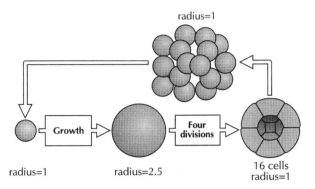

radius=1

Growth

radius=1

radius=2.5

Four divisions

16 cells
radius=1

FIGURE 6 The origin of the embryo derived by a cell increasing in size and then undergoing a number of divisions.

with nuclear division followed by the later establishment of cell boundaries. My own proposal is a modification of the second. It aims to provide a mechanism that accounts for the genetic identity of all the cells in the embryo and, crucially, the origin of the egg. The mechanism is based on the growth of a single cell well beyond its normal size followed by a sequence of divisions or cleavages.

Several changes are required in order for a single cell to give rise to a multicellular group by cell division. First, the cell has to grow larger than its normal size and this requires a transient block to mitosis. Secondly, in the enlarged cell the block to mitosis has to be overcome and the cell must then divide several times. And, thirdly, the cells have to remain together and separate only later.

The first two requirements involve modification of the cell cycle and it is relatively easy to propose mechanisms based on knowledge of controls in modern cells. During fission, yeast cell size at mitosis is enlarged in a good growth medium and reduced in a poor medium. It is therefore very possible that environmental cues could affect when the transition occurs. For example, a single cell growing in a good medium could become enlarged and on a shift to a poor medium undergo two divisions without further growth. This actually occurs with the protozoan *Chlamydomonas*: at fast growth rate cells divide into four or even eight at each cell cycle.

We can thus imagine a cell increasing its diameter 2.5-fold and then dividing four times to give sixteen cells (Figure 6). Let us further assume that the third requirement is satisfied and that the cells remain together, and, moreover, form a hollow sphere. This latter may require oriented cell divisions or the maintenance of junctions between cells on the outer surface but this does not seem to be too difficult to achieve.

The net result of such changes would be an increased cell size in good medium leading to a multicellular sphere in poor medium. It could be a selective advantage if the cells were ciliated so a sphere might swim faster and so find an environment of good food. Now there would be, for the first time, a positive selection for the multicellular state in a poor medium and the multicellular state would be environmentally induced.

Another 'just so story' for the origin of both multicellularity and the egg is based on cell death and cannibalism. Imagine a mutation so that, when a single cell divided, the two daughter cells remained together. This could have given this little animal a very important selective advantage over single-celled animals when food was in very short supply. While those that were just single cells might die of starvation, our two-celled ancestor could survive the hard times if the one cell ate the other. The surviving cell is the origin of the egg since in all multicellular organisms that develop from an egg the somatic cells sacrifice themselves for the survival of the egg. In many organisms there are examples of special cells whose sole function is to nurse the developing egg. So our origins may lie in cannabalistic altruism or altruistic suicide.

I now invoke a mechanism based on a process that is somewhat similar to the Baldwin effect which was proposed in the early years of this century and extended by the British embryologist Conrad Waddington into what he called 'genetic assimilation'. In essence it involves an environmentally produced effect becoming part of the developmental programme. An environmental signal is replaced by a developmental one. By the Baldwin effect we can imagine mutations such that, no matter what the medium is, the cells grow large and then divide many times. An environmental signal would have been taken over by the genes: the signal has, in a sense, become constitutive. The selective advantage for multicellularity could be speed of swimming, improved feeding or sharing of metabolites.

Just one more step is required for the evolution of the embryo. The individual cells needed to separate and start the programme again. This could have occurred when the cells grew and became too large to remain in contact. Each individual cell could now go through the same programme. Embryonic development had evolved. It could well have taken a long time to reach this simple, but crucial stage. In this primitive embryo, which we call a Blastaea, each cell behaves as an egg and there is no spatial organization. Given that the mechanisms of altered cell cycles could have generated a hollow sphere formed by a single sheet of cells, we can now consider the origins of patterning, the origins of spatial organization, which lead to cell specialization.

The key to all development is the generation of differences between the cells; that is, making them non-equivalent. Only if the cells are different can the organism be patterned so that there are organized changes in shape and cells at specific sites differentiate into different cell types. The alga *Volvox* illustrates some of these principles. The egg divides and gives rise to somatic and germ line cells, arranged essentially as a hollow sphere. The germ line cells are specified by unequal cleavages in which the daughter cells are of different size, the larger ones developing into germ cells.

How could mechanisms like those based on positional information have evolved? Consider that the primitive Blastaea came to rest on the ocean bottom (Figure 7). At the site where it made contact with the substratum it is not unreasonable to assume that these cells will be differentially affected. It could alter metabolism or affect surface receptors. In time this environmental signal could result in a cascade of activities starting at the point of contact, affecting both the cells at the contact site and signals which influence more distant cells. For example, there could be a selective advantage to the organisms becoming attached to the point of contact. Or it may be an advantage for the cells to invaginate at the site of contact. Whatever the advantage, an environmental signal could have brought about a localized change in the organism which becomes elaborated with time. It could even lead to suppression of growth of adjacent cells and so the restriction to reproduction of cells at the opposite end of the embryo. An embryonic axis could be specified. If the signal was graded it could have provided the origin of patterning based on positional information. Even today there are many such examples: light causing polarization of the egg of the algae *Fucus*; sperm entry determining the dorso-ventral axis in amphibians.

Using the concept of the Baldwin effect again, we can imagine how the specification of an environmentally induced axis could be incorporated into the developmental programme. All the machinery for specifying the axis was now present and all that was required was to replace the environmental signal by generating special cells in which the signal would be constitutive. One way of

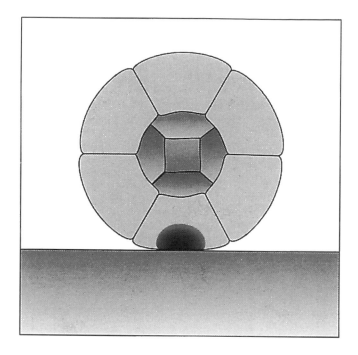

FIGURE 7 The possible origin of an axis at the point of contact between the Blastaea and the substratum.

doing this would be to have cytoplasmic localization in the egg which could define germ cells, which is a common phenomenon in modern-day animals.

Further evolutionary changes involved gene duplication which could generate new cells types, and mechanisms for generating periodic structures that led to segmented animals. Mechanisms for establishing positional identities using homeobox genes would have evolved early and together these processes provided a potent means of generating an enormous variety of multicellular organisms.

Gastrulation

While laying down the body plan involves patterning there are also very important changes in shape of the embryo, the most dramatic of which is gastrulation. During gastrulation cells which are on the outside of the embryo move inwards. This is necessary because the cells of the mesoderm and the

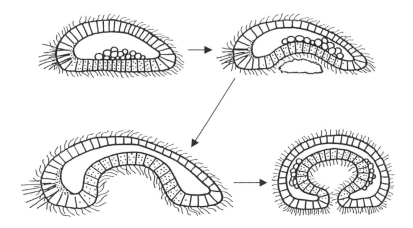

FIGURE 8   A possible origin of gastrulation. The Blastaea forms a primitive gut
in order to improve its ability to catch small organisms on which it feeds. This
resembles gastrulation in animals such as sea urchins when the gut forms.

endoderm – which give rise to the skeleton and the gut – are initially specified
in a layer in the outer regions of the embryo. They move to their appropriate
positions during gastrulation. The movements are rather complicated and can
best be visualized by letting a spherical balloon represent the early embryo. If
one places one's finger in at one point that could define the anus, then push
until the finger touches the other side, this is where the mouth will form and
the tube-like indentation is the gut. Gastrulation occurs in most animals and its
persistence may be due to the fact that it is much easier to pattern tissues in
two dimensions – like the spherical sheet of cells in the early embryo – and then
to create the third dimension by the sheet moving inwards. Such changes in
shape always follow an earlier patterning process, particular cells then generating the required forces.

The evolution of gastrulation transformed the embryo from a single-layered
system to a two-layered and then three-layered one. It was suggested in the last
century by the German biologist Ernst Haeckel that gastrulation had its origin
in a primitive organism, the Gastraea. The idea was that the even more primitive Blastaea, whose evolution we have just considered, may have settled on the
ocean bottom. A small infolding may have given a selective advantage if it
enabled the Blastaea to gather and ingest food more efficiently (Figure 8). This
invagination could have increased in size and become a primitive gut. It only

required for the tip of the invagination to fuse with the sheet on the other side – as in sea urchins – for a proper gut to have been established. Inward movement of individual cells could have provided the origin of the mesoderm. It is striking how many modern-day animals, like the sea urchin, gastrulate in a way that strongly resembles this primitive scenario.

Evolutionary privileges of the embryo

What selection pressures act on the developing embryo? Unlike the adult organism or larva, the embryo seems to be rather privileged. It need not, for example, seek food or mate, and so is not in direct competition for ecological niches. Its primary function is to develop reliably and this reliability is the main feature on which selection will act. This in no way excludes selection for aspects of development that relate more to reproduction and life cycles than to structure. These modifications include how much yolk is present, rate of development and the evolution of larval forms.

If the main selection pressure is for reliability there is the possibility that the embryo is evolutionarily privileged in that variations in development that do not affect reliability will not be subject to negative selection. For example, the expression of a gene in cells other than where it is required, the secretion of a variety of molecules or a transient invagination will escape negative selection provided it does not interfere with development. There is, for example, no evidence that such processes are energetically costly and so not subject to negative selection. It is, by contrast, striking how costly it is just to be alive, the sodium and calcium pumps alone using about 25% of total available energy and protein synthesis and breakdown another 50%. This lack of negative selection may offer the embryo the possibility to explore developmental pathways. The main selective presence is to develop reliably. Like pampered medical students, embryos can lead a dissolute life so long as, in the end, they pass their 'exams'.

Consider, for example, the primitive Blastaea or Gastraea before the mesoderm had evolved. Imagine a small number of cells moving into the interior by a chance mutation. The cells could continue to be present for many generations and so could themselves differentiate in a variety of ways. Some will be selected against, while others may, for example, generate internal muscles. The main selection pressure is on the adult. Again, the widespread secretion of signals and gene expression offer greater opportunities for useful combinations to emerge.

An important consequence of the privileged position of the embryo is that different pathways of development, leading to a similar end result, are to be expected. No one way of gastrulation is 'better' than another; only the end result matters. And there are, indeed, many variations, for example in the pathways leading to the simple two-layered early embryo of some invertebrates.

Conclusions

Given the eukaryotic cell it is not too difficult to imagine scenarios in which multicellular organisms develop by cleavage of a single cell. It is also possible to imagine how eggs became specific as distinct from body cells and how axes and patterning evolved. The evolution of these processes may have initially involved environmental signals which then became autonomous via the Baldwin effect. Gene duplication and segmentation involving duplication of structures, together with patterning based on positional identities and asymmetrical cell divisions, opened up pathways for the development of divergent cell patterns and structures.

Given these basic mechanisms it was possible to generate a very wide variety of different organisms. One may even think that these developmental mechanisms were selected because they could generate, relatively easily, variety. For example, a mechanism based on positional information dissociates the generation of differences from how the cells can develop. The same set of positional identities can generate many different patterns. This, together with the evolutionarily privileged status of the embryo, may have facilitated the development of increasingly complex organisms.

FURTHER READING

Akam, M., Holland, P., Ingham, P. and Wray, G. (eds.). 'The evolution of developmental mechanisms', *Development Supplement* (1994).

Lawrence, P. *The Making of a Fly*, Oxford: Blackwell, 1992.

Maynard Smith, J. and Szathmáry, E. *The Major Transitions in Evolution*, Oxford, New York, Heidelberg: W. H. Freeman/Spektrum, 1995.

Raff, R. A. *The Shape of Life: Genes Development and the Evolution of Animal Form*, Chicago, London: University of Chicago Press, 1996.

Waddington, C. H. *The Strategy of the Genes*, London: Allen & Unwin, 1957.

Wolpert, L. *The Triumph of the Embryo*, Oxford: Oxford University Press, 1992.

3 The Evolution of Guns and Germs

JARED DIAMOND

This chapter sets itself the modest task of explaining the broad pattern of history on all the continents for the last 13 000 years. Why did history take such different courses for peoples of different continents?

Eurasians, especially peoples of Europe and eastern Asia, have spread around the globe. They and their overseas descendants now dominate the modern world in wealth and power. Other peoples, including most Africans, survived and have thrown off European domination but remain far behind in wealth and power. Still other peoples, including the original inhabitants of Australia, the Americas and southern Africa, are no longer masters of their own lands but have been decimated, subjugated and even exterminated by European colonialists. Why did history turn out that way, instead of the opposite way? Why were American Indians, Africans and Aboriginal Australians not the ones who conquered or exterminated Europeans and Asians?

This question can easily be pushed back one step further. By the year A.D. 1500, the approximate year when Europe's overseas expansion was just beginning, peoples of the different continents already differed greatly in technology and political organization. Much of Eurasia and North Africa was occupied by Iron Age states and empires, some of them on the verge of industrialization. Two Native American peoples, the Incas and Aztecs, ruled over Stone Age or nearly Bronze Age empires. Parts of sub-Saharan Africa were divided among small indigenous Iron Age states or chiefdoms. All peoples of Australia, New Guinea and the Pacific Islands, and many peoples of the Americas and sub-Saharan Africa, lived as Stone Age farmers or hunter–gatherers.

Obviously, those differences as of A.D. 1500 were the immediate cause of the modern world's inequalities. Iron Age empires conquered or exterminated Stone Age tribes. But how did the world get to be the way that it was in the year A.D. 1500?

This question, too, can be pushed back a further step, with the help of written histories and archaeological discoveries. Until the end of the last Ice Age around 11 000 B.C., all humans on all continents were still living as Stone Age hunter–gatherers. Different rates of development on different continents, from 11 000 B.C. to A.D. 1500, were what produced the inequalities of A.D. 1500. While Aboriginal Australians and Native American peoples remained Stone Age hunter–gatherers, most Eurasian peoples and many peoples of the Americas and sub-Saharan Africa gradually developed agriculture, herding, metallurgy and complex political organization. Parts of Eurasia, and one area of the Americas, developed indigenous writing as well. But each of these new developments appeared earlier in Eurasia than elsewhere. For instance, mass production of copper tools was only beginning to spread in the South American Andes in the centuries before A.D. 1500, but was already spreading in parts of Eurasia 5000 years before that. The stone technology of Native Tasmanians in A.D. 1500 was simpler than that of Upper Palaeolithic Europe tens of thousands of years earlier.

Hence we can finally rephrase our question about the origin of the modern world's inequalities as follows. Why did human development proceed at such different rates on different continents for the last 13 000 years? Those differing rates constitute the broadest pattern of history, and the subject of this chapter.

To appreciate how non-obvious is the answer to this question, imagine that a historically minded intelligent being from outer space visited the earth 50 000 years ago. If that visitor had been asked to predict which continent's people would develop technology most rapidly, and who would conquer whom, what would the extraterrestrial have predicted? The visitor might well have answered 'Africa', because human history there had a six million year head start over history on the other continents. The visitor might also have reasonably predicted 'Australia', the continent with perhaps the earliest evidence of anatomically and behaviourally fully modern humans, and with by far the earliest evidence for human use of watercraft. The visitor would surely have written off Europe, where *Homo sapiens* still had not arrived as of 50 000 years ago. To that visitor, the state of the modern world would be unexpected. What were the reasons for the unexpected outcome?

Dismissing progress and IQ

At this point, readers may be beginning to wonder: is this chapter going to be a glorification of so-called progress? Will it be a justification of the status quo, with all its gross injustices? Will it be an apology for racism? I should therefore make two things clear at the outset.

First, I do not hold political and economic development to be an unmitigated good for the human species. It is debatable whether most people alive today are happier or healthier than most hunter–gatherers used to be. We today are certainly at more imminent risk of self-destruction than were our ancestors of 13 000 years ago. I merely want to examine the development of economic and political power without taking a position on whether it has been good for most of us.

Second, I want to make clear that this chapter is not about differences in IQ, and that it will not assert that Europeans are smarter than other peoples. Many Europeans tacitly assume this, even though they may have learned that it is no longer considered politically correct to say so in public. Technologically primitive peoples are often considered to be biologically primitive. It seems especially convincing that Aboriginal Australians and many New Guineans remained illiterate Stone Age tribal hunter–gatherers for 50 000 years, on a continent where Europeans, within a century of their arrival, apparently built a literate industrial food-producing modern state. Does that not prove that Europeans themselves are superior to Aboriginal Australians?

Of course it doesn't. Europeans did not develop literacy, food production and government in Australia; they imported it to there from the outside. Many psychologists, especially in the USA, have tried unsuccessfully to document IQ differences among different people. My own anecdotal perception, from my thirty years of work in New Guinea, is that New Guineans appear on the average considerably more intelligent than Europeans. On reflection, that outcome is unsurprising. Natural selection related to intelligence operates much more ruthlessly in traditional New Guinea societies than in politically organized Europe, so that New Guineans probably have an average genetic advantage. In addition, most European children today suffer from the crippling developmental disadvantage of spending much of their time being passively entertained by radio, TV and movies, while traditional New Guinea children spend all of their waking time talking or otherwise active with other children and adults. All psychological studies are unanimous about the role of childhood stimulation in promoting mental development, and about the irreversible mental stunting associated with reduced childhood stimulation. The same considerations apply more generally to other industrial peoples compared to other so-called technologically primitive peoples.

We therefore have to turn the usual racist assumption on its head. Instead of asking how industrial peoples came to be smarter, we must ask: why is it that modern Stone Age peoples, despite probably being genetically smarter and undoubtedly being developmentally advantaged, were nevertheless technologically outstripped and conquered by Eurasians?

For these broad patterns of history over whole continents, and over thousands of years, the explanation cannot involve accidental appearances of individual geniuses, such as Alexander the Great happening to be born in Macedonia rather than in what is now Mississippi. I shall show that the answer to the question about history's broadest pattern has nothing to do with differences among peoples themselves, but instead lies in differences among the biological and geographical environments in which different peoples found themselves.

Europe and the New World: proximate factors

As our first continental comparison, let us consider the collision of the Old World and the New World that began with Columbus's voyage in A.D. 1492, because the proximate factors involved in the outcome are well understood. I shall now give a brief summary of North American, South American, European and Asian history, including animal domestication, plant domestication and the evolution of infectious diseases!

Most of us are familiar with the stories of how a few hundred Spaniards under Hernan Cortes overthrew the Aztec Empire, and how another few hundred Spaniards under Francisco Pizarro overthrew the Inca Empire. The populations of each of those empires numbered millions, possibly tens of millions. At the Inca city of Cajamarca in modern Peru, when Pizarro captured the Inca Emperor Atahualpa in 1532, Pizarro's Spaniards consisted of only 62 soldiers on horseback plus 106 foot soldiers, while Atahualpa was leading an Inca army of about 40 000 soldiers.

Most of us are also familiar with the frequently gruesome details of how other Europeans conquered other parts of the New World. The result is that

Europeans came to settle and dominate most of the New World, while the Native American population declined drastically from its level as of A.D. 1492. Why did it happen that way? Why did it not happen that Montezuma or Atahualpa led the Aztecs or Incas to conquer Europe?

The *proximate* reasons are obvious. Invading Europeans had steel swords and guns, while Native Americans had only stone and wooden weapons. Just as elsewhere in the world, horses gave the invading Spaniards another big advantage in their conquests of the Incas and Aztecs. Horses had been playing a decisive role in military history ever since they were domesticated at around 4000 B.C. in the Ukraine. Horses revolutionized warfare in the eastern Mediterranean after 2000 B.C., later let the Huns and Mongols terrorize Europe and provided the military basis for the kingdoms emerging in West Africa around A.D. 1000. From prehistoric times until the First World War, the speed of attack and retreat that a horse permitted, the shock of its charge and the raised fighting platform that it provided left foot soldiers nearly helpless in the open. Steel swords, guns and horses were the military advantages that repeatedly enabled troops of a few dozen mounted Spaniards to defeat South American Indian armies numbering in the thousands.

Nevertheless, guns, steel swords and horses were not the sole proximate factors in the European conquest of the New World. The Indians killed in battle by guns and swords were far outnumbered by those killed at home by infectious diseases such as smallpox and measles. Those diseases were endemic in Europe, and Europeans had had time to develop both genetic and immune resistance to them, but Indians initially had no such resistance. Diseases that were introduced with the Europeans spread from one Indian tribe to another, far in advance of the Europeans themselves, and killed an estimated 95% of the New World's Indian population.

The role played by infectious diseases in the New World was duplicated in many other parts of the world. For instance, epidemic diseases brought by Europeans decimated Aboriginal Australians, the Khoisan populations of southern Africa and the populations of many Pacific islands. But there are also cases where diseases worked against Europeans: the infectious diseases endemic to tropical Africa, South-east Asia and New Guinea were the most important obstacles to European colonization of those areas.

Finally, there is still another set of proximate factors to be considered. How is it that Pizarro and Cortes reached the New World at all, before Aztec and

Inca conquistadores could reach Europe? That depended in the first instance on ships reliably capable of crossing oceans. Europeans had such ships, while the Aztecs and Incas did not. Those ships were backed by the political organization that enabled Spain and other European countries to finance, build, staff and equip the ships. Equally crucial was the role of writing in permitting the quick spread of accurate detailed information, including maps, sailing directions and accounts by earlier voyagers to motivate later explorers. Writing may also be relevant to what seems to us today the incredible naïveté that permitted Atahualpa to walk into Pizarro's trap and permitted Montezuma to mistake Cortes for a returning god. Since the Incas had no writing and the Aztecs had only a short tradition of writing, they did not inherit knowledge of thousands of years of written history. That may have left them less able to anticipate a wide range of human behaviour and dirty tricks, and made Pizarro and Cortes better able to do so.

Europe and the New World: ultimate factors

So far, we have identified a series of proximate factors behind European colonization of the New World: ships, political organization and writing that brought Europeans to the New World; European germs that killed most Indians before they could reach the battle field; and guns, steel swords and horses that gave Europeans a big advantage on the battle field. Now, let us try to push the chain of causation back further. Why did these proximate advantages go to the Old World rather than to the New World? Theoretically, American Indians might have been the ones to develop steel swords and guns first, to develop ocean-going ships and empires and writing first, to be mounted on domestic animals more terrifying than horses and to bear germs worse than smallpox.

The part of that question that is easiest to answer concerns the reasons why Eurasia evolved the nastiest germs. It is striking that American Indians evolved no devastating epidemic diseases to give to Europeans, in return for the many devastating epidemic diseases that they received from the Old World.

There are two straightforward reasons for this gross imbalance. First, most of our familiar epidemic diseases can sustain themselves only in large dense human populations concentrated into villages and cities, which arose much earlier in the Old World than in the New World. Second, most human epidemic diseases evolved from similar epidemic diseases of the domestic animals with which we came into close contact. For example, measles arose from a disease of our cattle, influenza from a disease of pigs, smallpox from a disease of cows and falciparum malaria from a disease of birds such as chickens. The Americas had a very few native domesticated animal species from which humans could acquire diseases: just the llama/alpaca (varieties of the same ancestral species) and guinea pig in the Andes, the Muscovy duck in tropical South America, the turkey in Mexico and the dog throughout the Americas. In contrast, think of all the domesticated animal species native to Eurasia: the horse, cow, sheep, goat, pig and dog throughout Eurasia; many local domesticates, such as water buffalo and reindeer; many domesticated small mammals, such as cats and rabbits; and many domesticated birds, including chickens, geese and mallard ducks.

Let us now push the chain of reasoning back one step further. Why were there far more species of domesticated animals in Eurasia than in the Americas? Since the Americas harbour over a thousand native wild mammal species and several thousand wild bird species, you might initially suppose that the Americas offered plenty of starting material for domestication.

In fact, only a tiny fraction of wild mammal and bird species has been successfully domesticated, because domestication requires that a wild animal fulfil many prerequisites: a diet that humans can supply, a sufficiently rapid growth rate, willingness to breed in captivity, tractable disposition, a social structure involving submissive behaviour towards dominant members of the same species (a behaviour transferrable to dominant humans) and lack of a tendency to panic when fenced. Thousands of years ago, humans domesticated every possible large wild mammal species worth domesticating, with the result that there have been no significant additions in modern times, despite the efforts of modern science.

Eurasia ended up with the most domesticated animal species in part because it is the world's largest land mass and offered the most wild species to begin with. That pre-existing difference was magnified 13 000 years ago at the end of the last Ice Age, when more than 80% of the large mammal species of North and South America became extinct, probably exterminated by the first arriving Indians. Those extinctions included several species that might have furnished useful domesticated animals had they survived, such as North American horses and camels. As a result, American Indians inherited far fewer species of big wild mammals than did Eurasians, leaving them only with the llama/alpaca as a domesticate. Differences between the Old and New Worlds in domesticated

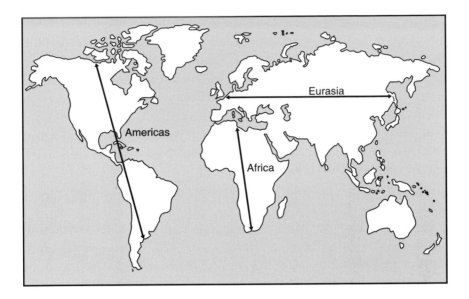

FIGURE 1 The major axis of Eurasia is east/west, facilitating rapid diffusion of crops and livestock and migration of peoples over long distances without encountering different latitudes, daylengths, climates or diseases. For the Americas and for Africa, the major axis is instead north/south, slowing diffusion and migration over long distances because of the need to adapt to different latitudes, daylengths, climates and diseases.

plants are qualitatively similar to these differences in domesticated mammals, though the difference is not so extreme.

A further reason for the higher local diversity of domesticated plants and animals in Eurasia than in the Americas is that Eurasia's main axis is east/west, whereas the main axis of the Americas is north/south (Figure 1). Eurasia's east/west axis meant that species domesticated in one part of Eurasia could easily spread thousands of miles at the same latitude, encountering the same daylength and climate to which they were already adapted. As a result, chickens and citrus fruit domesticated in South-east Asia quickly spread westwards to Europe, horses domesticated in the Ukraine quickly spread eastwards to China and the sheep, goats, cattle, wheat and barley of the Middle East quickly spread both west and east.

In contrast, the north/south axis of the Americas meant that species domesticated in one area could not spread far without encountering daylengths and climates to which they were not adapted. As a result, the turkey never spread from Mexico to the Andes; llamas/alpacas never spread from the Andes to Mexico, so that the Indian civilizations of Central and North America remained entirely without pack animals; and it took thousands of years for the corn that evolved in Mexico's climate to become modified into a corn adapted to the shorter growing season and seasonally changing daylength of North America. That seems to be the main reason why North America's Mississippi Valley, which you might think should have been fertile enough to support a populous and politically advanced Indian society, did not give rise to one until around A.D. 1000, when a variety of corn adapted to temperate latitudes was finally developed.

Eurasia's domesticated plants and animals were important for several other reasons besides letting Europeans develop nasty germs. Domesticated plants and animals yield far more calories per acre than do wild habitats, in which most species are inedible to humans. As a result, populations of farmers and herders are typically ten to 100 times greater than those of hunter–gatherers. That fact alone explains why farmers and herders almost everywhere in the world have been able to push hunter–gatherers out of land suitable for farming and herding. Domestic animals revolutionized land transport. They also revolutionized agriculture, by letting one farmer plough and manure much more land than the farmer could till or manure by his/her own efforts. In addition, hunter–gatherer societies tend to be egalitarian and have no political organization beyond the level of the band or tribe, whereas the food surpluses and storage made possible by agriculture permitted the development of stratified societies with political élites. The food surpluses produced by farmers also accelerated the development of technology, by supporting craftspeople who did not raise their own food and could instead devote themselves to developing metallurgy, writing, swords and guns.

Those professional specialists supported by agriculture also included full-time soldiers. That gave a decisive military advantage to many colonizing empires. For example, it was the decisive factor in the eventual success of New Zealand's British colonists at defeating New Zealand's indigenous Maori population, who were tough and well-armed fighters. While the Maori won some stunning temporary victories, each Maori man could fight for only a short time before having to go home to tend his garden. The Maori were eventually worn down by the full-time soldiers of the British colonists.

Thus, we began by identifying a series of proximate explanations – guns, germs and so on – for the conquest of the Americas by Europeans. Those proximate factors seem to me ultimately traceable in large part to the Old World's greater number of domesticated plants, much greater number of domesticated animals and east/west axis. The chain of causation is most direct in explaining the Old World's advantages of horses and nasty germs. But domesticated plants and animals also led more indirectly to Eurasia's advantage in guns, swords, ocean-going ships, political organization and writing, all of which were products of the large, dense, sedentary, stratified societies made possible by agriculture.

The history of Africa

Let us next examine whether this scheme, derived from the collision of Europeans with Native Americans, helps us to understand the broadest pattern of African history. I shall concentrate on the history of sub-Saharan Africa, because it was much more isolated from Eurasia by distance and climate than was North Africa, whose history is closely linked to Eurasia's history.

There are two big puzzles in the broad pattern of sub-Saharan African history. First, just as we asked why Cortes invaded Mexico before Montezuma could invade Europe, we can similarly ask why European countries colonized sub-Saharan African before sub-Saharan countries could colonize Europe. The promixate factors were the same familiar ones of guns, steel, ocean-going ships, political organization and writing; horses played much less of a role in Africa, and diseases in Africa may even have worked against Europeans rather than for them. Again, we can ask why guns and ships and so on ended up being developed in Europe rather than in sub-Saharan Africa. To the student of human evolution, this question is particularly puzzling, because humans have been evolving for millions of years longer in Africa than in Europe, and even anatomically modern *Homo sapiens* may have reached Europe from Africa only within the last 50 000 years. If time were a critical factor in the development of human societies, Africa should have enjoyed an enormous advantage over Europe.

The other puzzle in the broad pattern of African history is a collision within Africa. Until about 2000 years ago, most of sub-equatorial Africa seems to have been occupied by two groups of hunter-gatherers: pygmies in the moist equatorial areas, and Khoisan populations (alias Bushmen and Hottentots) throughout the drier parts of southern Africa. About 2000 years ago, Bantu populations originating ultimately from tropical West Africa rapidly expanded over almost the whole of southern Africa and replaced its Khoisan populations, except in the Cape Region and in dry areas unsuitable for agriculture. That Bantu expansion was powered by the advantages that Bantu gained over pygmies and Khoisan by possessing agriculture, herding and metal. Again, though, one can ask: why were those advantages developed by the Bantu rather than by the Khoisan?

Again, those advantages largely reflect biogeographical differences in the availability of domesticatable wild animal and plant species. Beginning with domestic animals, it is striking that the sole animal domesticated within sub-Saharan Africa was a bird, the guinea fowl. All of Africa's mammalian domesticates – cattle, sheep, goats, horses, even dogs – entered sub-Saharan Africa from the north, from Eurasia. At first that sounds astonishing, since we now think of Africa as *the* continent *par excellence* of big wild mammals. In fact, none of those famous big wild mammal species of Africa proved domesticatable. They were all eliminated by one or another problems such as unsuitable social organization, intractable behaviour, slow growth rate and so on. Imagine what the course of world history might have been like if Africa's rhinoceroses and hippopotamuses had lent themselves to domestication! Cavalry mounted on horses would have been helpless against cavalry mounted on rhinos or hippos. If those animals could have been domesticated, sub-Saharan Africans would have made mincemeat of Europeans. But it did not happen.

Instead, as I mentioned, the livestock adopted in Africa were Eurasian species that came in from the north. Africa's long axis, like that of the Americas, is north/south rather than east/west. Those domestic mammals spread southwards only slowly in Africa, because they had to adapt to different climate zones and different animal diseases. While cattle, sheep and goats reached the northern edge of the Serengeti Plains soon after 3000 B.C., it then took another 2000 years for them to cross the Serengeti and reach the Khoisan in southern Africa, just ahead of the invading Bantu.

The situation with domesticated plants in Africa is even more interesting. Agriculture based on indigenous wild plants did arise independently in Africa, from the equator *north* to the Sahara. Among those African domesticated plants, the one most familiar to readers of these pages is coffee, which was indigenous to Ethiopia, was domesticated there and has now spread around the world. Other plants domesticated between the Sahara and the equator include sorghum, several types of millet and yams and the oil palm. But no wild plant species was domesticated in Africa *south* of the equator. The result was that the Bantu developed agriculture but the Khoisan never did. Farming was instead carried into southern Africa by the invading Bantu, who were thereby able to displace most of the indigenous Khoisan peoples.

The difficulties posed by a north/south axis to the spread of domesticated species are even more striking for African crops than they are for livestock. Remember that the food staples of ancient Egypt were Fertile Crescent and Mediterranean crops like wheat and barley, which require winter rains and seasonal variation in daylength for their germination. Those crops were unable to spread south in Africa beyond Ethiopia, where the rains come in the summer and there is little or no seasonal variation in daylength. Instead, the development of agriculture in the Sahara and sub-Sahara had to await the domestication of native plant species like sorghum and millet, adapted to Central Africa's summer rains and relatively constant daylength.

Ironically, those crops of Central Africa were for the same reason then unable to spread south to the Mediterranean zone of South Africa, where once again winter rains and big seasonal variations in daylength prevailed. Instead, agriculture in South Africa's Cape region required crops adapted to winter rains and seasonally varying daylength, like the crops of the Fertile Crescent and Mediterranean. But those crops could not survive conditions in Central Africa and so could not be transmitted overland through chains of farmers from the Mediterranean to the Cape. Instead, those Mediterranean crops reached Africa's Cape region only with European settlers in the seventeenth century. The Bantu advance southwards halted in Natal, beyond which the zone of winter rainfall began and Bantu crops were unable to grow. Those facts about adaptations of domesticated plants had notorious consequences for modern South African politics, because Bantu farmers were not occupying the Cape when European farmers arrived.

A further consequence of Africa's north/south axis has to do with an irony of agriculture in modern tropical Africa. Some of modern tropical Africa's most important crops are no longer the crops native to tropical Africa, but are instead tropical Asian crops such as bananas, yams and taro, or tropical American crops such as corn and cassava. Because tropical Africa is flanked by oceans on both sides, tropical Asian crops did not reach Africa until Arab and

Indonesian traders began arriving across the Indian Ocean about 2000 years ago, while tropical American crops did not reach Africa until Europeans colonized the New World and then brought New World crops to Africa. If the Indian or Atlantic Oceans had been bridged by land similar to the broad east/west expanse of Eurasia, those productive tropical Asian and American crops would have reached tropical Africa thousands of years earlier, just as Asian chickens and citrus fruits had reached Europe.

In short, a north/south axis and a paucity of wild plant and animal species suitable for domestication were doubly decisive in African history, just as they were in Native American history. First, the indigenous Khoisan people of most of sub-equatorial Africa never developed nor adopted agriculture, and they acquired livestock from the north late, just before most of the Khoisan were overwhelmed by the far more numerous, better armed, Iron Age Bantu. Second, although the Bantu themselves had some plants domesticated locally in tropical West Africa, they acquired valuable domestic animals only later, from the north. The resulting advantages of Europeans in guns, ships, political organization and writing permitted Europeans to colonize Africa, rather than Africans to colonize Europe.

The history of Australia

Let us now conclude our whirlwind tour around the world by devoting some space to the last continent, Australia. In modern times, Australia was the sole continent still inhabited only by hunter–gatherers. Native Australia had no farmers or herders, no writing, no manufacture of metal tools and no political organization beyond the level of the tribe or band. Those are the reasons why European guns and germs destroyed Aboriginal Australian society. But why had all Native Australians remained hunter–gatherers?

There are three obvious reasons. First, even to this day no native Australian animal species and only one plant species (the macadamia nut) has proved suitable for domestication. There still are no domestic kangaroos.

Second, Australia is the smallest continent, and most of it can support only small human populations because of low rainfall and productivity. Hence the total number of Australian hunter–gatherers was only about 300 000.

Finally, Australia is the most isolated continent. The sole outside contacts of Aboriginal Australians were tenuous overwater ones with New Guineans and Indonesians. The coast of north-west Australia, where occasional visiting

Indonesian fishermen landed, is one of the most barren parts of Australia, quite unsuitable for growing any crops that the Indonesians might have brought with them. As a result, the cultural barrier between Australia and Indonesia or New Guinea remains astonishingly sharp. For example, in New Guinea there were bows and arrows, agriculture, pigs, chickens and pottery for thousands of years, but not one of those cultural items crossed the hundred-mile or so water gap of the Torres Strait to become established in Australia.

To get an idea of the significance of that small population size and isolation for the pace of development in Australia, consider the Australian island of Tasmania, which had the most extraordinary human society in the modern world. Tasmania is an island of about 26 000 square miles, lying 130 miles south of Australia at the latitude of Vladivostok or Chicago. When first visited by Europeans in 1642, Tasmania was occupied by 4000 hunter–gatherers related to mainland Australians, but with the simplest technology of any recent people on Earth. Cultural features that mainland Aboriginal Australians possessed and that Tasmanians lacked included the following. Tasmanians could not start a fire; if a family's fire went out, they had to get fire from neighbours to re-light it. The sole Tasmanian weapons were hand-held spears and clubs. They lacked the boomerangs, spear-throwers and shields of mainland Australians. Tasmanians had no bone tools, no specialized stone tools and no compound tools like an axe-head mounted on a handle. Their only stone tool was a crude hand-held scraper without ground edges. With only those scrapers, Tasmanians could not fell a tree or hollow out a canoe. They lacked sewing, nets, traps and ropes. Since they could not sew, their clothing consisted just of a one-piece cape, occasionally worn by being thrown over the shoulder. Their watercraft were rafts that remained afloat for only about ten miles. Though they lived mostly on the sea coast, the Tasmanians – incredibly – did not catch or eat fish.

Anthropologists feel uncomfortable discussing the Tasmanians, because of the awful end of their society: they were exterminated by British settlers within a few decades. To acknowledge that the Tasmanians had a relatively simple technology *seems* to be construed as justifying their extermination, which is, of course, nonsense. In particular, some anthropologists argue that Tasmanians had simple technology because they did not need anything more complicated. That interpretation is also manifestly incorrect. For humans anywhere in the world, it is convenient to be able to light a fire, to have nets and traps, to be able to sew clothes in order to keep warm during cold wet winters and to have devices such as bows or spear-throwers in order to discharge a projectile with much greater force than one can discharge a hand-held spear. Tasmanians surely would have profited from those things as did all other peoples, and there is a real problem in explaining their absence in Tasmania. How did those gaps in Tasmanian material culture arise?

Remember that Tasmania used to be joined to the southern Australian mainland at Pleistocene times of low sea level, until the land bridge was severed by rising sea level 12 000 years ago. People walked out to Tasmania tens of thousands of years ago, when it was still part of Australia. Once that land bridge was severed, though, there was absolutely no further contact of Tasmanians with mainland Australians or with any other people until the Dutch explorer Abel Tasman arrived in 1642, because both Tasmanians and mainland Australians lacked watercraft capable of crossing those 130-mile straits between Tasmania and Australia. Tasmanian history is thus a study of human isolation unprecedented except in science fiction – namely, complete isolation from all other humans for 12 000 years.

If all those technologies that I mentioned, absent from Tasmania but present on the opposite Australian mainland, were invented by Australians within the last 12 000 years, we can surely conclude that the Tasmanians did not invent them independently. Astonishingly, the archaeological record demonstrates something further: Tasmanians actually *abandoned* some technologies that they brought with them from Australia and that persisted on the Australian mainland. For example, bone tools and the practice of fishing were both present in Tasmania at the time that the land bridge was severed, and both disappeared from Tasmania around 1500 B.C. That represents the loss of valuable technologies: fish could have been smoked to provide a winter food supply, and bone needles could have been used to sew warm clothes. What sense can we make of these cultural losses?

The only interpretation that makes sense to me goes as follows. All human societies go through fads in which they temporarily either adopt practices of little use or else abandon practices of considerable use. For example, there are several instances of people on Pacific islands suddenly deciding to taboo and kill off all of their pigs, even though pigs are their only big edible land mammal! Eventually, those Pacific islanders realize that pigs are useful after all, and they import a new breeding stock from another island. Whenever such senseless taboos arise in an area with many competing human societies, only some societies will adopt the taboo at a given time. Other societies will retain the useful practice, and will either outcompete the societies that lost it or else will be there as a model for the societies with the taboos to repent their error and reacquire the practice. If Tasmanians had remained in contact with mainland Australians, they could have rediscovered the value and technique of fishing and making bone tools. But that could not happen in the complete isolation of Tasmania, where cultural losses became irreversible.

In case it is still difficult for you to believe that these cultural losses really happened in Tasmania, there are similar examples from other Pacific islands, such as the isolated Chatham Islands east of New Zealand, settled by New Zealand Maori who proceeded to live there in complete isolation for at least five centuries. There are also fourteen small and isolated Pacific Islands on which human populations actually went extinct after many centuries. The best known of these mystery islands is Pitcairn, famous for its rediscovery by the *H.M.S. Bounty* mutineers many centuries after the disappearance of Pitcairn's former Polynesian population. All of those islands on which human populations actually disappeared were so small that they could have held at most a few hundred people. Evidently, a few hundred people are just too few to maintain human society indefinitely in total isolation. If so, the 4000 Tasmanians and the 2000 Chatham Islanders were enough to keep their societies alive, but not enough to protect their societies against significant cultural losses.

In short, the message of the differences between Tasmanian and mainland Australian society seems to be the following. All other things being equal, the rate of human invention is faster, and the rate of cultural loss slower, in areas occupied by many competing societies with many individuals and in contact with societies elsewhere. If this interpretation is correct, it is likely to be of much broader significance. It probably provides part of the explanation (in addition to Australia's paucity of domesticatable wild animal and plant species) for native Australians remaining as Stone Age hunter gatherers, while people of other continents were adopting agriculture and metal. It is also likely to contribute to the differences that I have discussed between the farmers of sub-Saharan Africa, of the much larger Americas and of the still larger Eurasia.

Conclusion and outlook

As for the overall meaning of this whirlwind tour through human history, it is that our history has been moulded by our environment. The broadest pattern of human history – namely, the differences between human societies on different continents – seems to me to be attributable to differences in continental environments. In particular, the availability of wild plant and animal species suitable for domestication, and the ease with which those species could spread without encountering unsuitable climates, have contributed decisively to the varying rates of rise of agriculture and herding, which in turn have contributed decisively to human population numbers, population densities and food surpluses, which in turn contributed decisively to the development of writing, technology and political organization. In addition, the histories of Tasmania and other isolated societies warn us that continental areas and isolations, by determining the number of competing societies, may have been another important factor in human development.

As a biologist also at home in laboratory experimental science, I am aware that these interpretations may be dismissed as unprovable speculation, because they are not founded on replicated laboratory experiments. The same objection can be raised against any of the historical sciences, including astronomy, evolutionary biology, geology and palaeontology. It can, of course, be raised against the entire field of history. That is the reason why we are uncomfortable about considering history as a science: it is classified as a social science, which is not considered quite scientific.

But remember that the word 'science' is not derived from the Latin word for 'replicated laboratory experiment', but instead from the Latin word for 'knowledge'. In science, we seek knowledge and understanding by whatever means are available and appropriate. There are many fields that no one hesitates to consider sciences, even though replicated laboratory experiments in those fields would be immoral, illegal or impossible. We cannot manipulate stars while maintaining other stars as controls, nor can we start and stop ice ages, nor can we experiment with evolving dinosaurs. Nevertheless, we can still gain considerable insight into these historical fields by other means. We should surely be able, then, to understand human history, since introspection and preserved writings give us far more insights into the ways of past humans than those of dinosaurs. For that reason I am optimistic that we can eventually arrive at convincing explanations for these broadest patterns of human history.

FURTHER READING

Cavalli-Sforza, L. L. and Cavalli-Sforza, F. *The Great Human Diasporas*, Reading, MA: Addison-Wesley, 1995.

Crosby, A. *Ecological Imperialism: The Biological Expansion of Europe, 900–1900*, Cambridge: Cambridge University Press, 1986.

Diamond, J. M. *Guns, Germs, and Steel*, New York: W. W. Norton, London: Jonathan Cape/Random House, 1997.

Zohary, D. and Hopf, M. *Domestication of Plants in the Old World*, 2nd edition, Oxford: Oxford University Press, 1993.

4   The Evolution of London

RICHARD ROGERS

If evolution is the process of cumulative change then cities can be characterized as metabolisms that adapt over time in order to survive. In this chapter I argue that London, like other complex urban systems, is a fragile and delicate structure that has come full circle in the cycle of evolutionary change. The city is now facing an environmental, social and political crisis that threatens its very existence. Its future cannot be left to processes of random mutation, where market forces determine the policies and shape of its physical fabric. Intelligent forward planning, anticipatory design and government intervention are necessary to avoid the process of gradual decline and eventual extinction that has affected urban cultures in the past.

As the magnitude of the global environmental crisis comes to light, the design and management of our cities (the major consumers of world energy and producers of pollution) are becoming more critical to the survival of the planet. The solution lies in cities such as London becoming sustainable – circular metabolisms that conserve resources, use renewable energies and recycle waste, providing an equitable and stable social environment for future generations.

Sustainable growth

London is not alone in the struggle for urban survival. Human life has always depended on three variables: population, resources and environment. Today, we are the first generation of human inhabitants of the planet to face *and* be aware of the simultaneous impact of expanding populations, depletion of resources and erosion of the environment. All this is common knowledge, and yet, industrial expansion and urban growth carries on regardless. London's environmental crisis is all the more poignant because it is emblematic of the postindustrial city facing the consequences of unchecked economic growth and exploitation of natural and artificial resources.

Historically, societies unable to solve their environmental crisis have either migrated or perished. The vital difference today is that the scale of our crisis is no longer regional but global. It involves the entire planet. Worldwide communications bring home – to the citizens of Mexico City, Detroit or London alike – the scale, immediacy and impact of the environmental crisis. It is becoming increasingly clear that the same parameters that drive environmental decline are generating disastrous social instability. Social and environmental issues are interlocked. In the face of such overwhelming negative evidence, it is easy to forget that cities, which are now failing to provide the most basic needs of society, can provide a healthy and civilizing environment for our citizens. It is for this reason that the design of our cities has never been so crucial.

As well as being the greatest consumers of world energy, cities and buildings are the most rapacious destroyers of the ecosystem. In London, massive traffic congestion causes more air pollution today than there was before the Clean Air Act banned the burning of coal in 1956. Foul air is blamed for the fact that one in seven city children in Britain now suffers from asthma (and as many as one in ten in certain Eastern European cities). In the USA pollution rising from the cities has reduced crop production by 5–10%. Tokyo dumps an estimated twenty million tons of waste every year. The city has already saturated its bay with waste, and is now running out of sites on land. London, therefore, is only one of the major international cities in decline. But its unique history, wealth and human resources place it in a strong position to bring about a radical change in environmental consciousness, a change founded on the concept of sustainable development.

In recognition of the magnitude of the environmental crisis and its socio-economic impact on the equilibrium of the planet, the United Nations laid down the concept of 'sustainable development' as the backbone of global economic policy (the Brundtland Report, *Our Common Future*, 1987). Its aim is that we should try to meet our present needs without compromising the existence of future generations. At the core of this concept is a new notion of 'wealth' that incorporates those environmental elements previously considered limitless and free: clean air, fresh water, an effective ozone layer and a fertile land and sea. The ultimate aim of sustainable economic development is to leave to future generations a stock of environmental wealth, or nature capital, that equals or exceeds our own inheritance.

Nowhere is the implementation of 'sustainability' more relevant than in cities, particularly in those developed countries which have always provided a model (good and bad) for the poorer developing world. London could provide a new model of sustainable development that becomes the guiding law of modern urban design. Once again, London could have a radical impact on the city of the future as it did on the development of the nineteenth-century city at the peak of the Industrial Revolution.

The single-minded city

When it comes to cities, sustainability must be defined in social and cultural terms as well as environmental and economic ones. Poverty, unemployment, ill-health, poor education – in short, social injustice – all undermine a city's capacity to carry out these policies. In this respect London is beginning to resemble the stereotypical image of the modern city: a fortress city, polarized into ghettos of rich and poor, segregated into single-use zones – the business district, the housing estate, the residential suburb – with large monofunctional buildings, leisure complexes and shopping centres, destroying the delicate mixed-use texture of the traditional core of the city.

The American political theorist Michael Walzer had defined the structural difference in terms of 'single-minded' and 'open-minded' spaces. The first are designed by planners and developers with only one purpose in view. The second cater for a variety of uses in which everyone can participate. Both types of space have a role to play in the city. Single-minded spaces cater to our very modern craving for private consumption, for autonomy and intimacy. By contrast, open-minded places give us something in common: they bring diverse sections of society together, and breed a sense of tolerance, identity and mutual respect.

Open-mindedness in cities is giving way to single-mindedness. This is not just an American urban phenomenon. In London, local businesses and shops have been drained from the high street in Shoreditch, Limehouse or Chiswick, and removed to commercial centres or business parks along the major motorways, particularly the M25. Traditional mixed-use communities across London are being replaced by strictly segregated residential enclaves, industrial parks and commercial centres. Recently, the Department of the Environment admitted that government planning policy on out-of-town shopping centres has caused the ruin of our market towns and inner cities.

London's towns and villages have suffered. Small businesses in the inner city are weighed down by high rents and business rates, and polluted and congested streets. The inevitable outcome is that the streets, squares and other components of the public domain lose the very diversity that creates their animation. As public spaces decline, the natural policing of streets and sense of security that comes from the very presence of people is lost. Public spaces become little more than functional areas – highways for scurrying pedestrians or sealed, private cars. As London becomes increasingly divided into ghettos of activity, the spaces between them become increasingly soulless. Walk through a housing estate in Hoxton, Tower Hamlets or Dalston and you can feel the intensity of alienation. The Broadwater Farm and Brixton riots of the late 1980s are the social corollary of this level of urban and environmental degradation. They are tangible symptoms of the same malaise that has brought three civil uprisings to Los Angeles in the last three decades, a devastating indictment of our inability to control our destiny in the so-called developed world. Despite its humanist façade, London is in danger of going the same way.

London: an evolutionary metabolism

But why has London, still revered as one of the most attractive and desirable cities in the Western world, reached this critical moment in its evolution? It has, in effect, undergone a sequence of periods of concerted growth and radical change that parallels the evolutionary cycle of living organisms. From a small provincial outpost of the Roman Empire, London grew into the first megalopolis of the Industrial Revolution – the most congested, polluted and unhealthy city in the world. The City of Dreadful Night, 'reeking with poisonous and malodorous gases [where] thousands of beings herd together . . . in rotten and reeking tenements', according to Andrew Mearns (1883). London was transformed, through intervention, into a civilized city. The birth of the London County Council (LCC) and the adoption of pioneering planning legislation set the standard for dense urban living around the world. With its urban parks, public housing, modern sewers, street lighting and an effective public transport system grafted onto a city on the verge of explosion, London became the model of the modern humanist city.

But the urban organism, fuelled by a rampant economy, continued to grow. In the first four decades of the twentieth century the population expanded from five to eight million. The problems of congestion and inadequate housing were compounded by severe bombing during the Second World War. But at the height of war the English architect and pioneer of town planning Patrick

Abercrombie drafted the Greater London Plan that was to redirect London's growth in a fundamental way. The plan comprised four concentric rings, with the inner core at its centre and the countryside at its outer fringes. The Green Belt was conceived of as a *cordon sanitaire* to contain the city and protect the countryside. Population growth was absorbed in a constellation of new satellite towns – Harlow, Basildon, Stevenage and Hatfield – linked by a major radial route, more or less in the position of the M25 orbital motorway.

As an urban masterplan Abercrombie's plan ensured the city's survival. It checked uncontrolled urban sprawl (which could have led to endless miles of American-like suburbs) and channelled growth and development into strategically placed new towns. The plan allowed London to absorb the extraordinary pressures of growth resulting from the economic expansion and the population explosion of the 1950s and 1960s, and – to a degree – has been able to cope with the onslaught of the motorcar until the 1970s.

London in decline, 1975–1995

Since the 1970s, though, London has entered a new period of intensive change, but this time one of steady decline with no controlling mechanisms in place. In fact, this decline has coincided with the death of strategic planning following the vindictive destruction (rather than reformation) of the Greater London Council by the central government. London, Europe's largest city, no longer has a strategic planning authority. Instead its responsibilities now fall between five government departments, thirty-three London boroughs, the City of London and some sixty committees and 'quangos'. The London Planning Advisory Council, whose sole mandate is to offer 'advice' to the government, is isolated in Romford thirteen miles from the heart of the capital.

The city is spreading outwards in ever increasing circles. It is now served by a commuter belt 200 miles wide from Cambridge to Southampton, the largest and most complex urban region in Europe. London is both sprawling *and* emptying out at the centre. This pattern of development has produced a city of huge contrasts. One of the richest cities in the world has seven of the ten most deprived boroughs in the country at its centre, most of them in the east. As the city spreads, the rot sets in at the core. Already 5% of inner London is derelict, including large sites in areas such as Wandsworth, Vauxhall, Greenwich, Shepherd's Bush, Lambeth, Hoxton, Waterloo and King's Cross. This figure excludes the huge tracts of redundant industrial land along the Thames in

London's Docklands which were once the life blood of the city's mercantile economy.

Docklands: lost opportunity or a sustainable future

With the demise of river-borne shipping and the creation of a container port at Tilbury in the 1970s, the fifteen-mile corridor of land leading eastwards from London to the coast became the major opportunity for growth and consolidation of the capital. It is in this direction that Britain will become permanently linked to Europe through the international rail network. The Isle of Dogs, Surrey Docks and the Royal Docks provide ideal opportunities to create genuinely vibrant and sustainable new communities. The availability of land, the presence of water, views across the river and the prospects of good international transport, are the ingredients for mixed and balanced communities.

But government urban policy since the early 1980s has effectively excluded any strategic planning approach that encourages new mixed communities. The *laissez-faire*, market-led approach (the closest approximation to a process of random mutation) has produced a hybrid city, broken up into chaotic zones of commercial development, clumps of offices here, clusters of housing there, with no vision or civic quality. With the creation of London Docklands Development Corporation and the Enterprise Zone, central government set about wooing the private investor with massive tax reliefs. Instead of gaining a vibrant and humane new borough tied into the framework of the larger metropolis, Londoners have got a clutch of indifferent commercial buildings linked by an 'amusement park' train. The taxpayer lost money and the community was alienated from the decision-making process, promised everything and left only with big buildings and more traffic.

If the right lessons can be drawn from the Docklands experience, there is a chance of creating new opportunities for London's future through sustainable growth. The new train link to Europe provides particular hope. The impact that this will have on that part of London can hardly be exaggerated. Fifty years ago, 'Heath Row' was an unknown village, now it is the busiest international airport in the world which has generated wealth, employment and development to the west of London. With vision and strategic planning the rail connection and its urban spin-offs could have a similarly beneficial effect on East London and the Thames corridor, structuring growth polycentrically around ecologically and socially sustainable towns and community centres.

Traffic, congestion and public transport

But London needs to resolve a number of fundamental infrastructure issues before it can tackle the redevelopment of specific sites and contemplate future growth. As the major generator of urban sprawl and pollution, the car has led the onslaught on the city throughout the twentieth century. Just as the elevator made the skyscraper possible, so the car has enabled citizens to drive away from polluted city centres. The car has made viable the concept of dividing our everyday activities into compartments, segregated zones of offices, shops and homes. And the wider our cities spread out, the more it becomes uneconomic to expand public transport systems and the less they can offer an adequate alternative to driving.

Yet the paradox is that the car is perhaps the century's most liberating and most desired technological product. To the individual it is cheap, practical and promises freedom and status. In most cities of the world, however, it is cars that are now generating the bulk of air pollution, the very same pollution that the city dwellers are fleeing. Between 1970 and 1995, car ownership in Europe alone more than doubled. It is continuing to increase in developed cities and is about to soar in developing ones.

In London, traffic congestion is a critical issue. Fear of traffic has a profound effect on the way we behave. Parents are not allowing young children to cross the road on their own, and this is isolating children from their friends, and making them less independent. Nationwide, over the past twenty-five years, the number of seven-to-eight year-olds going to school on their own has fallen from 80% to 9%. Concern over road safety is encouraging parents to move away from the centre of London.

Transport lies at the heart of any strategy for making our cities sustainable. Today, London has no concerted transport policy. Instead its transport is dominated by private-car use which is polluting the metropolis, and undermining its communities. A staggering two-thirds of journeys within London are by car. The government itself predicts that there will be a rise of 142% in vehicular traffic over the next twenty-five years. But whilst £3 billion of taxpayers' money was spent on roadworks in Britain in 1994, major public transport initiatives have been shunned. Compare a 1930s map of the London Underground with a contemporary edition, and you will see that they are basically the same.

But the government is showing few signs of determination even though it accepts that road transport is the major source of air pollution in Britain. Pollution is contributing to one in seven children in London suffering from asthma, or more severe respiratory disease. In winter 1994 record traffic pollution was blamed for 155 deaths in just four days in London. And if pollution were not bad enough, the Confederation of British Industry estimates that traffic congestion cost London £10 billion in 1989 alone in wasted energy and time.

Average ticket prices for buses, rail and underground in London are approximately twice those in Paris, and five times more than Madrid. With the belated exception of the Jubilee Line Extension, new railway or underground lines such as the Chelsea–Hackney line or Crossrail are delayed or simply abandoned.

Other cities are addressing their problems of congestion and pollution with far more determination, vision and courage. London needs to implement policies to reduce pollution and congestion, and improve public transport. Tax benefits should be given to those who purchase small engine cars with catalytic converters or, better still, electric cars. A system of road pricing should be implemented to deter cross-town traffic. Research predicts that a 30% reduction could be easily achieved if road pricing were introduced in London today.

But these policies could undermine mobility, and especially that of the poor, unless public transport is improved, and made more affordable. London needs a co-ordinated transport strategy that has been evaluated in economic, ecological and social terms. Currently, car travel is cheap because it is sponsored by the taxpayer. The external costs of driving – road building, maintenance, subsidies for business cars, pollution, corrosion of buildings (currently a massive 3% of Gross National Product), disruption to the local community and ill health – are simply not reflected in the cost of car, road tax or petrol tax.

Public expenditure on public transport is argued to be unjustifiable on economic grounds. Yet there are many economists, of all political persuasions, who reject this notion recognizing the fact that society must differentiate between notions of short-term accounting and of long-term investment. Public transport infrastructure is a long-term investment. London, more than any other world city, is still reaping the benefits of (private) investments made in the city's underground, rail and bus system more than a century ago. Investment in a decent, clean and efficient public transport system will be useful to society for decades, possibly centuries to come. Its cost must be viewed in the long term, and in the context of the overall improvements to the city as a whole. Good public transport makes London more competitive and energy efficient and Londoners more mobile and healthy. It will make their city more beautiful.

Public spaces or traffic roundabouts

The car has not only affected the life style, economy and environment of London, it has had a negative impact on its physical infrastructure, contributing substantially to the erosion of the public realm. London's great public spaces – Parliament Square, Piccadilly Circus, Trafalgar Square – and its court-yards such as Horse Guard's Parade, Somerset House and Burlington House have been overwhelmed by cars. And this situation actually gets worse in local centres, such as Hammersmith, Shepherd's Bush, Brixton, Dalston or the Elephant and Castle. A shift from private to public transport will give Londoners the opportunity to swap highways for public places.

Trafalgar Square, once the heart of the Empire, is now the centre of a round-about. Tourists are disappointed by it; Londoners ignore it. But the Square could easily be transformed by turning the road which cuts it off from the National Gallery into a pedestrian precinct. A new gyratory at the top of White-hall would allow this transformation. This would create a new open-air sculp-ture terrace, linking the National Gallery to the Square, and liberating the Square to the rest of London with cafés and activities placed in a discreet arcade below the existing terrace. Like the Campo in Siena or the Place Beau-bourg in Paris, Trafalgar Square would become a vibrant urbane meeting place – somewhere to enjoy outstanding views of the turrets, domes and towers of Whitehall, and the Palace of Westminster.

Parliament Square might also regain its stature linked to Westminster Abbey. Shepherd's Bush could forge a link with its surrounding community. These spaces could be made into civic squares for people, outdoor living rooms for London's citizens.

At the macro scale, the same concept of urban linkage could be applied to the wider network of London's local parks and green areas. These reservoirs of public amenities could be easily linked together along quieter routes. Where no links exist, canal ways and towpaths could be used, abandoned railway lines could be landscaped or traffic-calming measures could be implemented together with pedestrianization of strategic routes and places. It would not be

FIGURE 1 A view of London along the Thames.

hard, for instance, to create continuous cycle routes from Richmond Park or Clapham Common – with new pedestrian bridges – to Hampstead Heath and beyond.

The Thames: an underused resource

But the real heart of London is the river. Look at any satellite image and it is the Thames that dominates. Today the river separates the poorer South of London from the more prosperous North (Figure 1). It is this huge and beautiful waterway which holds the key to revitalizing the metropolis. It has the potential of acting as a single cohesive element that links together diverse and segregated communities. An effective river transport system would make the Thames a stepping stone from which to reach the rest of London. Transport piers from

Kew to Greenwich could be integrated into London's transport network animating the heart of new communities, with shops, and amenities – forming the basis of a linear city focused on the Thames.

The stretch of the river from Westminster Bridge to Tower Bridge, at the very centre of the nation's capital, is an underused public amenity, bordering on London's richest and poorest communities. This area should become the focus of a great Millennium Project – a project about public places rather than National Monuments. The banks which ring this stretch of the river contain some of our most famous buildings, and some of our most important cultural institutions – from the Houses of Parliament to the Tower of London, from Tower Bridge to the Festival Hall. It is also less than 500 yards from Covent Garden, St Paul's, the Strand, the Old Vic and Waterloo Station, although we never associate these places with the river.

A new riverside park could stretch from Parliament to Blackfriars on the north bank of the Thames by re-routing the busy embankment motorway. It would consolidate the existing gardens to create a sweeping three-mile long riverside park with cafés and restaurants, the first major park built in London in the twentieth century. The park could also spill out into the river, onto floating islands with moored ships, pontoon and continuous boardwalks linking major monuments. Central London has only one-third of the number of bridges compared to central Paris. It could be more integrated as a city, with more pedestrian crossings feeding into the natural flow of movement across the city.

If many of these concepts appear utopian or unrealistic on economic grounds, one should consider the extraordinary opportunity for London's built (as opposed to natural) environment provided by the National Lottery and Millennium Fund. Most of these funds will be spent on public buildings – museums, theatres and sports centres, stimulating urban revitalization and providing the opportunity for the citizen to participate in the revival of British architecture. But it is the spaces between all these institutions and buildings, the public domain of London, that should benefit from this windfall. The National Lottery should be perceived as a once-in-a-lifetime chance to think big, to resolve London's problems at a macro scale, rather than patching together worthy though piecemeal proposals for individual buildings and urban areas. A special case, to my mind, must be made for the transformation of the Thames into a vital public element of London's life, drawing life back to the very heart of the capital.

A compact urban strategy

Many of these environmental issues are common to cities worldwide. But unlike London, other cities such as Barcelona, San Francisco and Rotterdam are implementing positive action as a way of controlling the next stage of urban evolution. These cities recognize that designing the physical form of new or existing communities has a precise social and environmental impact. It is here that the notion of sustainable development is most relevant as an evolutionary principle. London could easily build upon its existing physical structure to form a new network of dense, diverse and compact environments that reverse the trend of inner city degradation, 'ghettoization' and urban sprawl.

As Abercrombie's Green Belt plan has confirmed, a fundamental benefit of the compact approach is that the countryside is protected from the encroachment of urban development. For the citizen, the compactness is enhanced by a network of open public spaces and parks at the core of the inner city. The compact city grows around a number of centres of social and commercial activity. These are the focus around which communities develop into neighbourhoods. London's historic structure of towns, villages and parks is typical of this polycentric pattern.

The advantages of the compact strategy is that the city's neighbourhoods – with their own parks and public spaces – combine a diversity of private activities with public services and facilities. They generate local employment opportunities within convenient reach of the community. This proximity means less driving around for everyday needs, and makes trams, light rail, electric buses, cycling and walking more pleasant and more effective. Congestion and pollution from cars is reduced whilst the potential for random encounter and urban liveliness is enhanced.

Metropolitan Public Transport Systems, such as the Paris RER, providing high-speed cross-town travel by linking one neighbourhood centre to another, should be encouraged in London, leaving local distribution to local systems. This would also reduce the impact of through-traffic which should be calmed and controlled, especially around the public heart of neighbourhoods. Proximity (and its consequent reduction in congestion), the presence of landscape and the exploitation of new urban technologies can radically impove air quality in the city. The concentration of buildings can make for more efficient distribution and recycling of power.

Sustainable development in London and other cities could stimulate the implementation of community-based urban society. Cities are about people, about face-to-face contact and the expression of local culture. Whether the climate is hot or cold, the community rich or poor, the long-term aim of sustainable development is to create a flexible structure for a vigorous community with a healthy and non-polluting environment.

The huge scale of modern international cities and the sheer power of modern technology have led us away from historical practices of sustaining cities within their environment. Modern-day city building works at such a level of complexity – both technical and human – and at such a prodigious pace, that it has simply been impossible to co-ordinate a sustainable approach. The city continues to be seen as a machine to be exploited, rather than part of an environment to be nurtured.

We do not have to wait for the opportunity of starting from scratch to create a truly sustainable environment. Sustainable approaches can most readily be applied to urban renewal. Like London, most cities of the developed world have suffered intense de-industrialization over the past twenty years, leaving a legacy of unemployment and vast areas of derelict and abandoned land. Much of this land – like London's Docklands – is in, or very close to, the centres of cities, offering tremendous possibilities for sustainable development close to sources of energy, transport and local communities.

Tackling the social problems of our cities involves a radical re-thinking of collective values. But even here, the recent revolution in attitudes to the natural environment paves the way. Just as the natural environment is being re-evaluated as an indispensable resource, the built environment should be assessed according to the same parameters. The terms with which ecologists describe our relationship to nature – the idea that we are not owners but trustees, with duties to future generations – apply just as well to the city's public life.

In this light it makes no sense to give London's Embankment over to a motorway, when we would all enjoy a riverside park so much more. Perhaps the very best reason for respecting and promoting public life is simply that it is immensely pleasurable. Public life fosters tolerance and a sense of community. It is no coincidence that in racist or fascist societies the city was segregated. Sharing public space forces us to acknowledge what we have in common.

London: the basis for a sustainable growth

London's inherent civic heritage – a constellation of towns and villages, from Hampstead to Westminster, Notting Hill to Limehouse – should be used to consolidate the city, around sustainable, diverse and compact urban neighbourhoods. The fact that London is made up of a collection of distinct towns and villages means that each community already benefits from its own sense of character, visual identity, history and community. Instead of letting London sprawl, and allowing these communities to decline, we should be actively reinforcing neighbourhoods. Large areas of dereliction should be used to establish new mixed and diverse communities.

The sustainable city of the future will need to be one of many facets. Despite its considerable problems, London offers the potential of becoming one of the first, truly sustainable, complex urban systems, a city composed of many interrelated ingredients. First, London should build upon its structure as a dense and polycentric city – because this form of settlement protects the countryside, focuses communities around neighbourhoods and minimizes dependence on cars. Secondly, London should encourage public access to, and mixed use of, new developments because it maximizes contact and diversity, and fosters a vital public life. Thirdly, London should once again set the international standard for the equitable city – a self-governing participatory city where wealth, justice and opportunity are fairly distributed. Fourthly, London should exploit the opportunity to become an ecological city with a circular metabolism which gives as much to the environment as it takes out. Fifthly, London should play the role of an open city which embraces new ideas and experiments with architectural form. Finally, London should consolidate its reputation as a beautiful city, where art, architecture and landscape move the spirit.

It is encouraging that current economic trends underpin this strategic approach as a viable economic option. The industrial age is giving way, at least in the developed countries, to the postindustrial age of telecommunications, cheap computer power, the information superhighway, clean robotic and microscopic manufacturing. All this has the potential to transform the character of our cities for the better. The raw material of this new economy is, as ever, citizens and their knowledge – creativity and initiative. Art and science will be the life blood of knowledge-based development and the key to creation of further wealth. Networks of small-scale companies are emerging as the driving force of the economy of the future.

New multimedia technologies and industries could help to overcome the barriers between urban zones of housing, offices and factories – indeed the distinctions between office and home, work and play, education and entertainment are themselves set to dissolve. This type of small-scale economy will give a city such as London a finer and more diverse texture of overlapping activities and neighbourhoods which will facilitate the emergence of a dynamic, greener, more community-based city. Yet the new communication industries will only flourish within cities that have the right mixture of educational facilities, public life and creative leadership. Barcelona, Glasgow and Lyons are already striking out on their own, doing everything to establish themselves as centres of the new, city-friendly communications age. That is why it is so important that we invest in London and other British cities, and provide them with the resources and independence they need to regain control of their evolution and destiny.

FURTHER READING

Girardet, H. *The Gaia Atlas of Cities*, London: Gaia Books, 1992.

Hall, P. *Cities of Tomorrow*, Oxford, Blackwells, 1992.

Lovelock, J. *The Ages of Guia*, Oxford: Oxford University Press, 1988.

Pearce, D. *Blueprint II, Greening of the World Economy*, London: Earthscan Publications Ltd, 1991.

Porritt, J. and Winner, D. *The Green History of the World*, London: Fontana, 1988.

Porter, R. *London, A Social History*, Cambridge, MA: Harvard University Press, 1995.

Rogers, R. *Cities for a Small Planet*, London: Faber and Faber, 1997.

Walzer, M. 'Pleasures and costs of urbanity', *Dissent Magazine*, Autumn (1986), 470–5.

5   The Evolution of Society

TIM INGOLD

Evolution

Many years ago I heard a lecture on evolution by a distinguished geneticist. Holding a stone in his hand, he observed that, were he to let it go, there was a fair degree of certainty that it would fall to the ground. With that, I am sure everyone in his audience agreed. It is equally certain, he then went on to declare, that species have evolved. This beguiling analogy has stuck in my mind ever since, for three reasons. First, declarations of certainty seem an odd place from which to start doing science. After all, it was only because Darwin refused to accept the certainty that species had been created to divine order that we have a theory of evolution at all. Secondly, I was put in mind of the objection lodged by Canon Kingsley, over a century ago, to the claim that there was a similar inevitability about the evolution of society. A dropped stone, Kingsley noted, would not necessarily hit the ground if someone decided to catch it. His point, of course, was that human freedom could not readily be comprehended within a framework of mechanical law. Thirdly, I was moved to reflect that had it not been for a colossal misunderstanding in the history of their subject, brought about through an uncritical extension of widely held ideas about social evolution to the organic domain, contemporary biologists would now be telling us that to believe that species have evolved is profoundly misguided. Let me explain.

The verb 'to evolve', from the Latin *evolvere*, originally meant to roll out or unfold. Darwin, as is well known, used the word only once in the first edition of *The Origin of Species*. It is, in fact, the very last word of the book, and is used in its original sense to convey the idea of the history of life as a grand procession of forms unfolding before the timeless gaze of the observing naturalist. As the earth has gone on rotating according to fixed gravitational laws, so, wrote Darwin, 'endless forms most beautiful and most wonderful have been, and are

being, evolved'. This is just one more metaphorical image – a concluding flourish – in a work that teems with such images. When it came to the actual changes
undergone by species, changes which he sought to explain by his theory of variation under natural selection, Darwin was a great deal more precise. He spoke
not of evolution but of 'descent with modification', meaning by that the sequential generation of genealogically connected forms, each minutely different from
those preceding and following. Indeed, he would have had good reason to have
avoided the concept of evolution. For having first entered biology with Charles
Bonnet's homunculus theory of embryonic preformation, the concept had just
been hijacked by the social philosopher Herbert Spencer in an altogether different sense, but one no less alien to the founding premises of Darwin's theory.

Spencer had learned, at second hand, about the work of the German
embryologist Karl Ernst Ritter von Baer, who had advanced the claim that the
development of any organism consisted of a process of structural differentiation leading, in Spencer's rendering, 'from an incoherent homogeneity to a
coherent heterogeneity'. In an article dating from 1857, two years before
Darwin published his epoch-making book, Spencer speculated that this principle of development might govern not only the constitution of living organisms
from their cells but also the constitution of societies from their individual members, of minds from the elements of consciousness and, indeed, of the entire
universe from the basic constituents of matter. Having originally called his
principle the 'law of progress', he quickly substituted 'evolution' for 'progress'
on the grounds that the latter term was too closely associated with theories of
exclusively *human* development. In Spencer's vision the progress of humanity
was but part and parcel of the overall advance of life, which in turn was integral
to the development of the cosmos as a whole. On subsequently reading Darwin,
Spencer was convinced that he had found independent confirmation, from
within the field of biology, of his evolutionary law. Indeed, Spencer never
thought of Darwin's work as anything other than accessory to his own synthetic
philosophy.

It appears that Darwin was none too impressed with Spencer's grandiose
style of philosophical speculation. Nevertheless, urged on by the co-discoverer
of natural selection, Alfred Russell Wallace, he was persuaded to take on Spencer's phrase, 'the survival of the fittest', as a possible alternative to 'natural
selection' in later editions of the *Origin*. He could not, however, bring himself
to accept that the modification of species through natural selection necessarily

entailed progress or advance in any absolute sense. According to his theory, organisms should adapt to whatever the prevailing conditions of life might be, and if in the process they had advanced in terms of structural differentiation or overall complexity, the reasons were to be found in the particular conditions, not in the general mechanism. At root, Darwin was simply not concerned with the evolution of life as Spencer had conceived of it – that is, as one phase of a cosmic movement that continually builds itself, through its own properties of dynamic self-organization, into ever novel and increasingly complex structures. His aim was much more modest: to account for the endless adjustment, remodelling and fine-tuning of those manifold and ingenious contrivances by which the current of life – 'having been originally breathed', as he put it, 'into a few forms or into one' – had been carried into every nook and cranny of the habitable world. It was Spencer, not Darwin, who saw in this process of adaptive modification the hand of evolution at work. And in so doing he initiated a confusion that has been perpetuated by generations of biologists right down to such architects of the 'modern synthesis' as Theodosius Dobzhansky and Julian Huxley.

It is instructive to speculate on what might have happened if this confusion had never arisen. In place of today's evolutionary biology, with its sometimes exaggerated claims to have produced nothing less than a complete explanation of life, we would have a narrower, less presumptuous branch of biological science dealing specifically with the mechanics of organic adaptation. Following in Darwin's footsteps, its practitioners would regard themselves as students of descent with modification, though in keeping with the spirit of modern times they would surely have got into the habit of abbreviating it to DWM. No doubt the new generation of DWM theorists would be anxious to correct anyone foolish enough to think that the adaptive modification of species amounted to a kind of evolution. For that, they would say, is to confuse phylogenetic change with ontogenetic development. Only the latter, underwritten as it is by a unique programme encoded in the organism's genetic endowment, has the character of the progressive unfolding of organized complexity to which the concept of evolution properly applies. Descent with modification does not follow a programme; rather it occurs because of imperfections in the mechanism whereby information is copied from one generation to the next. These ensure that, in the development of organisms, no two programmes are ever quite the same. When it comes to the vexed question of whether or not evolution occurs in the domain

once called 'superorganic' – though now more commonly known as 'sociocultu-
ral' – our DWM theorists would remain non-committal. They would not, however, display the marked intolerance that their real-life counterparts (that is,
modern evolutionary biologists) have shown towards social scientists who continue to associate the idea of evolution with that of a movement of progressive
development in culture or society. Far from taking these social scientists to task
for misunderstanding the very nature of evolution, DWM theorists would
observe that the methods and concepts of the Darwinian paradigm are applicable to the explanation of social and cultural change only to the extent that the
latter is *not* an evolutionary process.

To continue with this scenario: let us suppose that I had been asked to contribute a chapter on the evolution of society. You and I would surely expect, as
our predecessors of a century ago would have done, that the chapter would
address the claim that social life is characterized by an irreversible process of
growth and development, not unlike that of the individual organism. Convinced as I am that social life is integral to the overall movement of organic life,
rather than conducted on a superior level of existence, I might have presented
you with a proposal for a new sociobiological synthesis. The biology I would
draw on, however, would not come from DWM theory. It would come instead
from the work of developmental biologists who have set out to understand the
dynamics of morphogenesis – the process, cutting across the emergent inter-
face between organism and environment, wherein organic form is generated
and held in place. In an article published in 1991, I sketched out a proposal
along these lines, and I am still of the view that developmental biology, rather
than DWM theory in its current neo-Darwinian guise, is the most promising
place from which to start in the project of integrating biological and social science. But the problem goes rather deeper than that. It is not just a matter of
deciding whether social evolution should be likened, in the first instance, to a
process of ontogenetic development or of phylogenetic change. More fundamentally, we need to reconsider the premises on which the distinction
between ontogeny and phylogeny has classically been drawn. I return to this
problem in the final part of this chapter.

Now the real historical situation in which I find myself, in contrast to the
imaginary scenario I have just presented, is one in which the concept of evolution has been appropriated by neo-Darwinian biology to signify a process of
phylogenetic change through variation under natural selection. As I have been

invited to write on the evolution of society, I am sure you will expect me to address the claim that social form is likewise in some way a product of a selective process. I do not believe that this claim can be justified, and I intend to argue against it. It is not my purpose, however, to contend that we need one kind of theory for human beings and another for the rest of the animal kingdom, nor do I mean to come out in favour of Canon Kingsley to the effect that human beings, having become conscious of the laws of nature, are free to contravene them whenever they feel like it. To the contrary, I aim to show that a paradox of neo-Darwinian evolutionary biology is that it presumes, yet cannot comprehend, the historical process whereby certain humans came to be in a position to formulate it. Although Darwin could explain natural selection, natural selection cannot explain Darwin! If we are to seek an understanding that would re-embed human history within the overall continuum of organic life, as I believe we should, then we shall have to recast the whole way we think about evolution. In what follows, I suggest how this might be done. Before proceeding further, however, I should turn to consider the significance of the other key term in my title. Allow me, then, to digress for a while upon the theme of the evolution of 'society'.

Society

The word society comes from the Latin *societas*, and appeared in English for the first time in the fourteenth century. Its primary meaning was one of companionship, a sense retained in our contemporary notions of 'sociable' and 'sociability', with their connotations of friendship and intimacy. In short, society stood for the positive qualities of warmth, familiarity and trust in interpersonal, face-to-face relations, qualities also epitomized by the concept of community. Indeed, until the seventeenth century, the terms *societas* and *communitas* figured as virtual synonyms. The eighteenth century, however, saw the beginnings of a decisive shift in the meaning of 'society', towards a more general and abstract sense, ever further removed from the lived experience of human beings in their actual relationships. Initially, this new conception, of what was called 'civil society', was bound up with a direct challenge to the entrenched power structures and traditional hierarchical divisions of the absolutist state. Thus, the idea of civil society derived its significance, in the eighteenth century, from its *opposition* to state power, positing against the rigidly inegalitarian regime of the state an association of free and equal citizens, each entitled to pursue his or her private interests by contracting with other such individuals whenever it was to their mutual advantage. In this liberal and democratic view, society was modelled on the market, and social relations on market transactions: transient, self-interested, involving only an external compact rather than any deep and enduring interpersonal involvement. Society, according to this model, was but the aggregate of interindividual transactions.

Many eighteenth- and nineteenth-century commentators bemoaned what they saw as the loss of the sense of community entailed in the establishment of the order of civil society. Gone were the trust, companionship and familiarity that were regarded, perhaps romantically, as the trademarks of the traditional agrarian or peasant community, and in their place were the multiple, competing and antagonistic interests of bourgeois society. This was the source of Darwin's famous metaphor of the 'struggle for existence', and for Spencer's of the 'survival of the fittest'. One of the classic statements of the opposition between community and society is to be found in the work of the German sociologist Ferdinand Tönnies, entitled *Gemeinschaft und Gesellschaft*, published in 1887. *Gemeinschaft* is conventionally translated as 'community'; *Gesellschaft* as either 'society' or 'association'. 'The elementary fact of Gessellschaft', Tönnies wrote, 'is the act of exchange which presents itself in its purest form if it is thought of as performed by individuals who are alien to each other, having nothing in common with each other, and confront each other in an essentially antagonistic and even hostile manner.' Thus, by the time Tönnies was writing there had occurred a complete about-face in the meaning of society, from its original sense of familiarity and sociability to the opposite pole of mutual antagonism and hostility.

I have noted that the idea of civil society, as an aggregate of interactions among discrete and competing interests, was opposed not only to the earlier notion of *societas* as a field of sociability, but also to the established, hierarchical conventions of state authority. However, the nature of the distinction between society and the state, and of the relation between them, has remained highly contentious. The problem arises because civil society is not really possible *without* the state. As political theorists have recognized ever since Thomas Hobbes wrote his *Leviathan* in 1651, a society based on the free pursuit of self-interest can work harmoniously only if it is regulated, such that no one, in pursuing their interests, should infringe upon the liberty of others to do the same. The very existence of civil society was supposed to depend upon the establishment of the state, conceived as a mechanism designed to facilitate the smooth functioning of social life. But other theorists identified society itself with the institutions of regulation, and ultimately therefore with the state. For them, transactions motivated by self-interest rather than obligation were considered – like those of the market place – to be purely economic rather than properly social. Society, in their view, was coterminous with the domains of law and morality, and consisted in a framework of rules and obligations backed ultimately by sanctions vested in its highest authority. Here the concept of society is constituted by its opposition neither to the community nor to the state but to the *individual*. This was the fundamental premiss upon which Emile Durkheim, writing towards the close of the nineteenth century, set out his manifesto for what was then the new science of sociology; it was also the crux of his disagreement with Herbert Spencer. For Spencer there was no higher purpose in society beyond the desires of its individual constituents. Durkheim, to the contrary, maintained that the contact among individuals in society is not an exclusively external one, but that it gives rise to a certain interpenetration of minds out of which there emerges a consciousness of a higher order – collective rather than individual – that constrains and disciplines the pursuit of innate desires in the name of society as a whole.

Thus it is that the recent history of ideas has bequeathed to us three quite different and apparently contradictory notions of what a society is. All three are situated within a long and continuing controversy among Western philosophers, statesmen and reformers about the proper exercise of human rights and responsibilities. In this controversy, the particular meaning attached to 'society' has varied according to its opposition, alternately, to such notions as individual, community and state. Against the individual, society connotes a domain of external regulation – identified either with the state itself or, in polities lacking centralized administration, with comparable regulative institutions – serving to curb the spontaneous expression of private interests on behalf of public ideals of collective justice and harmony. In other contexts, however, especially those of emergent nationalism, society comes to represent the power of the people – as a real or imagined community bound by shared history, language and sentiment – *against* the impersonal and bureaucratic forces of the state. In yet other contexts, society stands *opposed* to community, connoting the

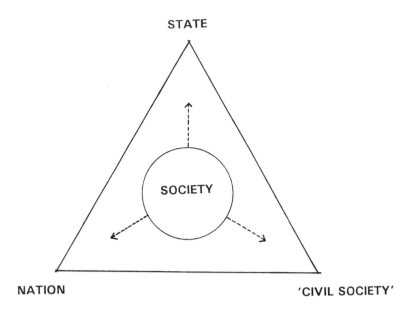

FIGURE 1 The meaning of society may vary within a semantic space defined by the ideals of 'civil society', the nation and the state.

mode of association of rational beings bound by contracts of mutual self-interest, as epitomized by the market, rather than by particularistic ties of the kind epitomized, for example, by relations of kinship or friendship.

To what, then, do we refer when we speak of, say, 'British society'? Perhaps we mean something rather closer to an imagined community of people than to a free association of citizens, yet closer to the government and institutions of the state than to the nation and, again, closer to an association of citizens than to the state. In short, the meaning of society may be pulled this way and that, as shown in Figure 1, within a triangular semantic space whose points are represented by the ideals of civil society, the nation as a community and the state as a sovereign authority. And from this brief excursus on the career of a concept, the conclusion to be drawn, which I take from a recent article entitled 'Inventing society' by the anthropologist Eric Wolf, is that assertions about the nature and existence of society are not simple statements of brute fact, but are rather *claims*, 'advanced and enacted in order to construct a state of affairs that previously was not'. In other words, the concept of society does not stand outside of all time and change, nor does it denote some eternal verity about the condition of humanity, as though societies were things that have always existed 'out there', independently of the claims that have, from time to time, been made about them. Rather, as Wolf points out, 'the concept of society has a history, a historical function within a determinate context, in a particular part of the world'. Different people, located at particular moments in this history, and with definite political purposes, have adopted the concept, shaped it to their require-ments and put it to use in order to advance their claims or lend substance to their cause.

But is not history itself a process of social life, carried on through the inten-tional activity of persons who are *already situated* in relationships and environ-mental contexts bequeathed to them by virtue of the actions of their predeces-sors? As Karl Marx wrote, in the *Eighteenth Brumaire* of 1869, 'men make their own history, but they do not make it just as they please, they do not make it under circumstances chosen by themselves, but under circumstances directly encountered, given and transmitted from the past'. Those who would seek to construct an order of society, on whatever ideological foundation, must already dwell in a world of other persons and relationships, so that the institutional forms they create are themselves constituted within the flow of social life. To put it another way, the reality of social life is no more *contained* within the things we call societies than is history contained within the fabrications of the human mind. To grasp this reality, Wolf advises us, we need to think *rela-tionally* – 'in terms of relations engendered, constructed, expanded, abrogated; in terms of intersects and overlaps, rather than in terms of solid, bounded, homogeneous entities that perjure without question and without change'.

By adopting such a relational perspective it becomes possible to see how there can be social life in the absence of anything recognizable as a society at all. To exemplify this point, I should like to refer, in a general way, to what numerous studies have shown concerning the form of life of people known to us as hunters and gatherers. It is characteristic of this form of life that people relate to one another, and indeed to components of the non-human environ-ment as well, on the basis of intimate familiarity and companionship. This is what is meant by the oft-stated observation that hunter–gatherer social life is founded upon face-to-face relationships. There is a sense of mutuality, of people relating to one another directly rather than as incumbents of offices or of formal status positions. This mutualism, however, is combined with a strong respect for personal autonomy. One may act with others in mind, in the hope and expectation that they will do likewise, but one cannot *force* a response. Any attempt to compromise another person's autonomy of action, by placing him or her under obligation or compulsion, represents a betrayal of trust and a negation of the relationship. Yet for the same reason, normal sociability will be extended to anyone who is prepared to show the kind of consideration and sensitivity to the needs of others that are of the essence of being a person. There is, as the anthropologist James Woodburn has osbserved, 'simply no basis for exclusion'. In short, the world for hunters and gatherers is not a socially segmented one, for it is constituted by relations of incorporation rather than exclusion, by virtue of which others are 'drawn in' rather than 'parcelled out'.

I hope I have said enough to show that we are dealing here with a form of sociality that is utterly incompatible with the concept of society, whether by that is meant the interlocking interests of 'civil society', the imagined community of the ethnic group or nation, or the regulative structures of the state. For one thing, the hunter–gatherer's claim to personal autonomy is the very opposite of the individualism implicated in the Western discourse on civil society. Whereas the latter posits the individual as a self-contained, rational agent, constituted independently and in advance of his or her entry into the public arena of social interaction, the autonomy of the hunter–gatherer is *relational*, in that a person's capacity to act on his or her own initiative emerges through a history of continuing involvement with others in contexts of joint, practical activity. For another thing, in a world where sociability is not confined by boundaries of exclusion, people do not define themselves as 'us' rather than 'them', or as members of this group rather than that, nor do they have a word to describe themselves as a collectivity apart from the generic word for 'persons'. This is why outsiders – explorers, traders, missionaries, anthropologists – seeking names with which to designate what they have perceived as discrete bands, tribes or societies of hunter–gatherers, have very often ended up borrowing exogenous labels applied as terms of abuse by neighbouring peoples towards the hunters and gatherers in their vicinity. Finally, the principle of trust which lies at the heart of hunter–gatherer sociality will not accept relations of domination of any kind. Yet such relations are necessarily entailed in any system of regulative institutions which legitimate and empower certain persons, in the name of society, to control the actions of others. It is therefore not enough to observe, in a now rather dated anthropological idiom, that hunter–gatherers live in 'stateless societies', as though their social lives were in some sense lacking or incomplete, waiting to be completed by the evolutionary development of a state apparatus. Rather, the principle of their sociality, as Pierre Clastres put it in the title of his 1974 book, is fundamentally *against* the state.

Science and the hunter–gatherer

Let me now return to the problem of evolution, while remaining for the moment with the anthropology of hunters and gatherers. For modern theorists of human evolution, hunter–gatherers have a very special significance – so special, indeed, that had they not existed they would almost certainly have had to have been invented. Evolutionary theory, it seems, *requires* hunter–gatherers. To show why, I shall have to raise the spectre of an old question that has exercised the minds of Western thinkers for centuries without, apparently, bringing us any closer to a resolution. This is the question of whether human beings differ from other animals in degree or in kind. The idea that no radical break separates the human species from the rest of the animal kingdom is an ancient one, going back to the classical doctrine that all creatures can be placed on a single scale of nature or Great Chain of Being, connecting the lowest to the highest forms of life in an unbroken sequence. Every step along the chain was conceived as a gradual one or, as the saying went, 'nature never makes leaps'. Darwin, in his theory of evolution by natural selection, replaced the image of the single chain with that of a branching tree, but the idea of gradual change remained. According to the view of the evolution of our species that you will find in any modern textbook, our ancestors became human by degrees, over countless generations. An unbroken sequence of forms is supposed to link the apes of some five million years ago, from which both human beings and chimpanzees are descended, through the earliest hominid creatures of two million years ago, to people like you and me – certified humans of the 'anatomically modern' variety: *Homo sapiens sapiens*.

Now, as an account of human biological evolution, that may be all very well, but what about human history? Theorists of the eighteenth century, cleaving to the philosophy of the Enlightenment, tended to think of history as the story of humanity's rise from primitive savagery to modern science and civilization. Yet they were also committed to the doctrine that all human beings, in all places and times, share a common set of basic intellectual capacities, and in that sense may be considered equal. This doctrine was known as the 'psychic unity of mankind'. Differences in levels of civilization were attributed to the unequal *development* of these common capacities. It was as though allegedly primitive peoples were at an earlier stage in their pursuit of a core curriculum common to humankind as a whole. In short, for these eighteenth-century thinkers, human beings differed in *degree* from other creatures with regard to their anatomical form, but nevertheless differed in *kind* from the rest of the animal kingdom in so far as they had been endowed with *minds* – that is, with the capacities of reason, imagination and language – which could undergo their own historical development within the framework of a constant bodily form. Even Linnaeus, who took the bold step of including human beings within his taxonomic system under the designation *Homo*, was hard pressed to discover any definite criteria by which to distinguish anatomically between humans and apes, choosing instead to identify the human distinction by means of a word of advice: *nosce te ipsum* – 'know for yourself'. Only humans, Linnaeus thought, could seek to know, through their own powers of observation and analysis, what kinds of beings they are. There are no scientists among the animals.

The immediate impact of Darwin's theory of human evolution, set out in his 1871 volume on *The Descent of Man*, was to subvert this distinction. Differences in mental capacity were attributed to different degrees of development of a bodily organ, the brain, such that civilized people were supposed to have larger, better organized brains than primitive people, just as the brains of the latter were supposed to be larger and better organized than those of the apes. Human history – or what had now come to be called the evolution of society – was understood to march hand-in-hand with the evolution of the brain, through a process of natural selection in which the hapless savage, cast in the role of the vanquished in the struggle for existence, was sooner or later destined for extinction. When Wallace suggested, in his *Contributions to the Theory of Natural Selection* of 1870, that the brains of primitive savages might be just as good as those of European philosophers, and therefore designed to be capable of more than was actually required of them under their simple conditions of life, he was dismissed as a spiritualist crank. For natural selection, it was argued, will furnish the savage only with as much brain power as he needs to get by. Only a Creator would come to think of preparing the savage for civilization in advance of his achieving it.

But, of course, Darwin was wrong and Wallace was right, although few give him credit for it. The brains of allegedly primitive hunter–gatherers *are* just as good, and just as capable of handling complex and sophisticated ideas, as the brains of Western scientists and philosophers. However, racist notions about the innate mental superiority of white European colonizers over indigenous peoples were remarkably persistent in biological anthropology. It was really not until after the Second World War, and the atrocities of the Holocaust, that such notions ceased to be tolerated in scientific circles. But this left the Darwinians with a problem on their hands. How was the doctrine of evolutionary continuity to be reconciled with the new-found commitment to universal human rights? The Declaration on Human Rights of the United Nations asserted, once again, the fundamental equality of all humans – present and future and, by implication, past as well. If all humans are alike in their possession of reason and moral conscience – if, in other words, all humans are the kinds of beings who, according to Western juridical precepts, can exercise rights and responsibilities – then they must differ in kind from all other beings which cannot. And somewhere along the line, our ancestors must have crossed a threshold from one condition to the other, from nature to humanity.

Faced with this problem, there was only one way for modern science to go; that is, back to the eighteenth century. Indeed, the majority of contemporary commentators on human evolution appear to be vigorously, if unwittingly, reproducing the eighteenth-century view in all its essentials. There is one process (evolution) leading from our ape-like ancestors to human beings of a biologically, or 'anatomically', modern form; another process (culture or history) leading from humanity's primitive past to modern science and civilization *while leaving us biologically unchanged*. History, as psychologists David Premack and Ann James Premack have recently pronounced, is 'the sequence of changes through which a species passes while remaining biologically stable', and of all the species in the world, only humans have it. Taken together, as shown in Figure 2, the axes of biological evolution and culture history establish by their intersection a unique point of origin, without precedent in the evolution of life, at which our ancestors are deemed to have crossed the threshold to true humanity and to have embarked upon the course of history.

Now it is a remarkable fact that whenever scientists are concerned to stress the evolutionary continuity between apes and humans, the humans are almost always portrayed as ancient hunter–gatherers (or if contemporary hunter–gatherers are taken as examples, they are commonly regarded as cultural fossils, frozen in time at the starting point of history). According to a now widely

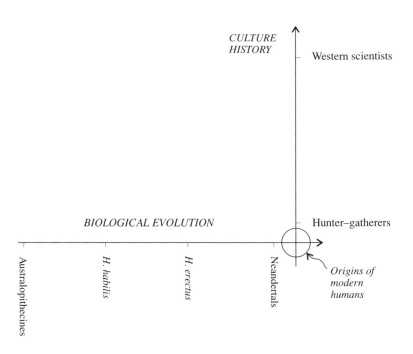

FIGURE 2 The origin of 'modern humans', at the point of intersection between the axes of biological evolution and culture history.

accepted scenario, it was under conditions of life as hunter–gatherers, in the Pleistocene era, that the biological capacities evolved – bipedalism, tool-use, big brains, male–female pair-bonding and so on – that are supposed to have made us human. Thus, every one of us is said to carry, as a fundamental part of our biological make-up, a set of capacities and dispositions that originally arose as adaptations to the requirements of hunting and gathering in Pleistocene environments. Of course, what was adaptively advantageous to our hunter–gatherer predecessors may not be so well suited to life in densely populated, urban environments, where people have access to high-powered technology whose potential destructiveness is vastly in excess of anything that our ancestors could possibly have envisaged. Many of the endemic problems of modern civilization, from road accidents to mechanized warfare, have been attributed to this. However, the idea that even the modern city dweller is afflicted by this legacy from our evolutionary past lies behind much of the continuing interest, both popular and academic, in contemporary hunters and gatherers whose form of life is thought to resemble most closely the condition of ancestral populations, and whose study might therefore reveal to us something of our inner nature. Inside each of us, it is supposed, there is a hunter–gatherer struggling to get out.

It should now be clear why Western thought and science, including the science of evolution, needs hunters and gatherers. In effect, the category 'hunter–gatherer' was brought in to characterize the original condition of humanity at the cross-roads of two processes of change – the one evolutionary, the other historical – whose separation is logically necessary in order to preserve the claim of science to deliver an authoritative account of the workings of nature in the face of the recognition that the scientist (who, like the rest of us, is only human) belongs to a species that has itself evolved to its present form through a process of variation under natural selection. Humans did not evolve as scientists, but they are thought to have evolved with the *capacity* to be scientists, and for that matter to read and write, to play the piano, drive cars and even fly rockets to the moon; indeed, to do anything that human beings have ever done or will do. Cro-Magnon man of 30 000 years ago, had he been brought up in the twentieth century, could have been an Einstein. His brain was as big, and as complex. But the time was not ripe, in his own era, for this potential to be 'brought out'. Stretched between the poles of nature and reason, epitomized respectively by the contrasting figures of the hunter–gatherer and the scientist, is supposed to lie the entirety of human history. There is a certain irony here. As I have already noted, biologists long ago co-opted the notion of evolution to describe what Darwin had called 'descent with modification', and have been scathing in their criticism of social scientists who have continued to use this notion in its original sense of progressive development. Yet they themselves cannot avoid a view of history – as the unfolding of pre-evolved potentials or capacities – that is fundamentally teleological!

In sum, contemporary evolutionary biology remains locked in the same contradiction that has been there all along. Its claim, that human beings differ from their predecessors in degree rather than kind, can be upheld only by attributing the total movement of history, from Pleistocene hunting and gathering to modern science and civilization, to a social or cultural process that differs in kind, not degree, from the process of evolution. This contradiction is, of course, but a specific instance of a more general paradox that lies at the heart of Western thought, which has no way of comprehending human beings' creative involvement in the world save by taking themselves out of it. The detachment or disengagement of the human observer from the world to be observed, to yield the dichotomy between reason and nature, is indeed central to the project of natural science, including the science of evolutionary biology. Gazing into the mirror of nature, the scientist sees his own powers of reason reflected back in the inverted form of natural selection. Despite the claims of evolutionary theorists to have dispensed with the archaic subject/object and mind/body dualisms of Western thought, they are still there, albeit displaced onto the opposition between the scientist, to whose sovereign imagination is revealed the design of nature, and the hunter–gatherer whose behaviour is interpreted as the output of innate dispositions installed by natural selection, and of which he or she has no conscious awareness. Even as neo-Darwinian biology proclaims the evolutionary continuity between humankind and the rest of the animal kingdom, it turns out that this continuity applies to humans as hunter–gatherers, not as scientists, and that the only way in which both scientists and hunter–gatherers can be brought within the same fold is by reasserting the essential distinction between humanity and nature, thereby compromising the thesis of continuity.

From evolution to history

To resolve the paradox of distinction and continuity, we need to find a mode of human understanding that starts from the premiss of our engagement with the world, rather than our detachment from it. This is what I take to be the central task of my own discipline of anthropology. And what makes anthropologists especially qualified to carry it out is their close familiarity with non-Western understandings. This is the point, then, at which I would like to return to my earlier discussion of the sociality of hunters and gatherers. I have shown that theirs is a sociality that is fundamentally *relational*, in the sense that persons come into being within the contexts of histories of continuing involvement with others. Relations are *enfolded* in persons, in their particular capacities, dispositions and identities, and *unfold* in purposive social action. This enfolding and unfolding, however, cannot be comprehended within the terms of the dominant Western discourse on the individual and society, a discourse that tends to deny to hunters and gatherers any social life at all. What happens if, instead of looking at hunter–gatherer lives through Western eyes we invert the perspective, and apply an understanding sharpened through listening to what hunters and gatherers have to tell us, to an examination of our own experience?

I believe we will find, then, that the underlying current of relational sociality is by no means limited to hunters and gatherers, but runs through and connects the lives of people everywhere, past and present, even modern urban dwellers like ourselves. If that is so, then the implications of this form of sociality as regards the constitution of persons can be generalized. It means that we can no longer accept the idea, central to neo-Darwinian orthodoxy, that human capacities are pre-specified, in advance of development, by virtue of some innate endowment that every individual receives at the point of conception. My contention, to the contrary, is that such capacities arise as emergent properties of the total developmental system constituted by way of the emplacement of the person-to-be, from the outset, within a wider field of relations – including, most importantly, relations with other persons.

I therefore differ from my colleague Michael Carrithers when he argues that sociality should be understood as an inherited, genetically encoded trait, 'expressed in individuals' and 'established through the force of natural selection'. For Carrithers, social relations are the manifest outcomes of the association of multiple individuals, each independently pre-programmed for co-operative or altruistic behaviour. My own view, by contrast, is that sociality is immanent in that very field of relations within which every human life is inaugurated and through which it seeks fulfilment. To be sure, there is a sense in which sociality is there from the start, and from that point of view it might be regarded as innate. By this, however, I mean that it is original to the constitution not of discrete individuals but of those relationships that comprise the dwelt-in world. Immanent in this world, sociality is the relational soil from which all human existence grows. Thus, instead of regarding it as a thing that evolves, we should think of sociality as the generative potential of a relational field, whose unfolding is tantamount to the evolutionary process itself. What, then, is the meaning of evolution?

To put it in the most general terms, evolution is the process in which organisms come into being with their particular forms and capacities and, through their environmentally situated actions, establish the conditions of development for their successors. Now human organisms are just as much caught up in this process as are organisms of non-human kinds. Human children, like the young of many other species, grow up in environments furnished by the work of previous generations, and as they do so they carry the forms of their dwelling in their bodies – in specific skills, sensibilities and dispositions. But they do not carry them in their genes, nor is it necessary to invoke some other kind of vehicle for the intergenerational transmission of information, cultural rather than genetic, to account for the diversity of human social arrangements. It is the very notion of information, that form is *brought in* to environmental contexts of development, that is at fault here. For as I have already shown, it is within the movement of social life, in the contexts of human beings' practical engagement with one another and with their non-human surroundings, that institutional forms – including those forms that go by the name of 'societies' – are generated.

Moreover, this movement, as we have seen, is none other than the process of history. I have already alluded to Marx's comment that history is something people make for themselves. Taking his cue from Marx, the anthropologist Maurice Godelier has proposed that humans make history because they do not merely live in society but play their part in creating it. My point, however, is that the creation of social forms does not take place in a vacuum, but against the background of the work people do, and have done in the past, in shaping the conditions of development for succeeding generations. Let me suggest an agricultural analogy. Farmers do not create crops; they *grow* them. Through their work in the fields, they establish the environmental conditions for the plants' healthy development. Now just as farmers grow crops, so people 'grow' one another. And it is in the growing of persons, I suggest, rather than in the creation of society, that history is made.

We can now see how, by taking the 'person in his/her environment', rather than the 'self-contained individual', as our point of departure, it is possible to dissolve the dichotomy between evolution and history which has been the source of so much trouble and misunderstanding in the past. As a movement in which people, through their own social practices in regard to one another, establish their respective conditions of development, history is but a specific instance of a process that is going on throughout the organic world. Hence, we do not need one theory to explain how apes became human, and another to explain how (some) humans became scientists. And once we come to recognize that history is the continuation of an evolutionary process by another name, the point of origin constituted by the intersection of evolutionary and historical axes disappears, and the search for the origins of society, history and true humanity becomes a search after an illusion.

But it is not only the division between evolution and history that falls with the argument I have proposed here. For it also strikes at the heart of the central principle on which orthodox theory distinguishes between evolution and development, or between phylogeny and ontogeny. The basis of this principle is that what every individual receives from its predecessors is a context-independent specification of form, known as the *genotype*, that is then expressed or 'realized', in the course of its life history, in the concrete form of an environmentally specific *phenotype*. Ever since the so-called Lamarckian doctrine of the inheritance of acquired characteristics was demolished by August Weismann, at the close of the nineteenth century, it has been assumed that only the characteristics of the genotype, and not those of the phenotype, are carried across generations.

That the constituent elements of design are thus imported into the organism, as a kind of evolved architecture, *prior* to the organism's development within an environmental context, is I believe one of the great delusions of modern biology. To be sure, every organism begins life with its complement of DNA in the genome, but on its own, DNA specifies nothing. There is no 'reading' of the genetic code that is not itself part of the organism's development in its environment. Of course, the organism does not begin life only with DNA. What is literally passed on from one generation to the next, as Susan Oyama has pointed out in her important book *The Ontogeny of Information: Developmental Systems and Evolution*, 'is a genome and a segment of the world'. Together these constitute a developmental system, and it is in the unfolding of this system, in the course of the life cycle of the organism, that form emerges and is sustained. Any account of the evolution of form must therefore be centrally concerned with the processes of dynamic self-organization whereby such systems are constituted and reconstituted over time. All I have done in this chapter is to establish the truth of this proposition for human organisms, growing up in a social world and playing their part in the making of history.

Let me conclude by returning to the image of the falling stone. I am not convinced that we can speak with anything like the same certainty about the evolution of organic form. I believe that the neo-Darwinian paradigm is riddled with contradictions, and I have tried to point some of these out. I would like to think, however, that Darwin himself, were he with us now, would look kindly on my endeavours. For Darwin was no Darwinist, let alone a neo-Darwinist, and he was a great deal more sensitive to the mutualism of organism and environment than many of those who nowadays yoke his name to their cause. But, above all, Darwin was a true scientist, who was prepared to challenge the orthodoxy of his time when reason, evidence and intellectual honesty required him to do so. It is curious, and not a little disturbing, that Darwin's heresy has now become a new orthodoxy, bordering in some cases almost on a faith. Those who claim that neo-Darwinism must be right *because there is no alternative*, and dismiss all doubters as heretics and enemies of Science, are surely the Wilberforces of the late twentieth century.

FURTHER READING

Carrithers, M. *Why Humans Have Cultures*, Oxford: Oxford University Press, 1992. [See, especially, Chapters 3 and 4. Carrithers argues that sociality is an innate, genetically transmitted trait of humans that has evolved through Darwinian natural selection.]

Clastres, P. *Society Against the State*, Oxford: Blackwell, 1977 (originally published in 1974 as *La Société contre l'état*). [Drawing on South American Indian ethnography, Clastres shows that the principles of their social and political organization are fundamentally contrary to those of the centralized state.]

Fortes, M. *Rules and the Emergence of Society* (Royal Anthropological Institute Occasional Paper 39), London: RAI, 1983. [In this brief, posthumously published work, the distinguished social anthropologist Meyer Fortes argues that human society is uniquely founded upon rules, and thus has no counterpart in the animal kingdom.]

Godelier, M. 'Incest taboo and the evolution of society', in *Evolution and its Influence*, ed. A. Grafen, pp. 63–92, Oxford: Clarendon Press, 1989. [Here, Maurice Godelier explores the implications of the thesis that human beings are the creators of their own societies, by looking at relations of kinship.]

Ingold, T. *Evolution and Social Life*, Cambridge: Cambridge University Press, 1986. [A study of the ways in which the idea of evolution has been handled in the context of anthropological debate, from the mid nineteenth century to the present day, comparing biological, historical and anthropological approaches to the study of human culture and social life.]

Ingold, T. 'Becoming persons: consciousness and sociality in human evolution', in *Evolutionary Models in the Social Sciences*, ed. T. Ingold (special issue of *Cultural Dynamics*, 4 (1991), 355–78). [I argue that personhood is not 'added on' to the human organism through socialization or enculturation, but rather arises within the process of the organism's development in an environment that includes, crucially, other organism-persons. Other articles in the same special issue, including those by Paul Graves, Mae-Wan Ho and John Shotter, explore related themes.]

Kuper, A. (ed.) *Conceptualising Society*, London: Routledge, 1992. [Several leading contemporary social and cultural anthropologists discuss what is meant by 'society' and 'sociality'.]

Oyama, S. *The Ontogeny of Information: Developmental Systems and Evolution*, Cambridge: Cambridge University Press, 1985. [A philosopher of biology shows how present-day thinking is still permeated by the nature/nurture dichotomy, and how overcoming this dichotomy requires a focus on the self-organizing properties of developmental systems.]

Premack, D. and Premack, A. J. 'Why animals have neither culture nor history', in *Companion Encyclopedia of Anthropology: Humanity, Culture and Social Life*, ed. T. Ingold, pp. 350–65, London: Routledge, 1994. [Comparing the different mechanisms by which information is passed on across the generations, Premack and Premack argue that humans are unique in their capacity to transmit knowledge through pedagogy, which in turn is the basis for both culture and history.]

Viveiros de Castro, E. 'Society', in *Encyclopedia of Social and Cultural Anthropology*, ed. A. Barnard and J. Spencer, pp. 514–22, London: Routledge, 1996. [A brilliantly succinct review of the different meanings of 'society' and their implications for anthropological theory.]

Wolf, E. 'Inventing society', *American Ethnologist* 15 (1988), 752–61. [Wolf explores the career of the concept of society in the recent history of Western ideas, and argues that the concept has now become an obstacle because of the way it predisposes us to think in terms of bounded units rather than fields of relationships.]

Woodburn, J. 'Egalitarian societies', *Man* (N.S.) 17 (1982), 431–51. [This article reviews ethnography on contemporary hunting and gathering societies, to show that, in certain of these societies, characterized by systems of production in which there is an immediate return on labour, equality is not only asserted in principle but also achieved in practice.]

6 The Evolution of the Novel

GILLIAN BEER

It seemed at first paradoxical to find myself invited to write about 'The Evolution of the Novel' since I do not believe that the novel evolved, and have spent some work demonstrating that this is one among a number of misplacements of the evolutionary metaphor – misplacements that have caused confusion (and worse) in other areas of experience (musicology and race-relations are two). But the invitation has allowed me to discriminate between, on the one hand, the effects of this misapplied metaphor on the history of past fiction and, on the other, the creative urgency with which writers have responded to evolutionary ideas in all their contradictory implications – indeed, *for* their contradictory implications. What has most drawn novelists, I argue, are the tensions within and between Darwinian ideas, as well as the pressures in the relation of such ideas to the human.

Fiction thrives always at points of contradiction, and evolutionary theory offered contradictory stories and contrasting trajectories for interpretation. Was this an account of development or decay? Did development inevitably imply progress, or was this a new version of the Fall? Did evolution make room for disinterestedness or did it necessitate always a ghastly struggle for too few resources? Was it communitarian or individualistic? What made evolutionary theory so fruitful for fiction was not only its coherence but its contradictions. True, it proposed a universalizing theory that would explain the history of all kinds on the earth: descent with modification, predominantly by means of the newly described and named principle, natural selection. The implications for human behaviour could be spelt out in quite incompatible ways; the more so since in *The Origin of Species* Darwin had suppressed the human figure altogether in his argument and in *The Descent of Man* was deeply influenced by 1860s ethnography and anthropology that was itself drawing largely on the

Origin to reinforce already and previously dominant theories of social development.

Some of the themes I shall be examining are limned lightly in *Tristram Shandy*, Laurence Sterne's novel whose first volume was published 100 years before the *Origin*, in 1759. Tristram Shandy's father is, as usual, orating; Uncle Toby is, as usual, trying to make sense of it:

> Kingdoms and provinces, and towns and cities, have they not their periods? and when these principles and powers, which at first cemented and put them together, have performed their several evolutions, they fall back. '—Brother Shandy, said my uncle Toby, laying down his pipe at the word *evolutions*— Revolutions. I meant, quoth my father—by heaven! I meant revolutions, brother Toby—evolutions is nonsense. —'Tis not nonsense—said my uncle Toby.'
>
> (*Tristram Shandy*, chap. 5, p. 3)

That contrast between *evolution* and *revolution* has taken on strong meanings since Sterne's time; but here Mr Shandy sees revolution as a wheel turning, and returning to the previous point. Evolution is the more radical term: 'evolutions is nonsense. 'Tis not nonsense—said my Uncle Toby.'

The passage I have quoted from *Shandy*, with its city states and government, its rise and fall of human societies, its progress and recession, brings out the degree to which human affairs are taken for granted as a model for the study of the natural world. (Karl Marx commented acutely in a letter to Friedrich Engels that 'in Darwin the animal kingdom figures as civil society'.) In the rhythmic biblical undersong of Sterne's exchange we hear already that Darwinian tone of threnody, in which the lost past is fleetingly glimpsed, altered, in the present – and in which the unknowableness of the future does *not* mean it is nonsense. Yet in Mr Shandy's argument and in the whole dialogue there is a suggestion of poise and settlement that Darwin will not share: 'they fall back', onto some foundation. That comfortable conservatism Darwin's stories never countenance: no originary state survives to which we can return.

To take first, briefly, the displacement of the evolutionary metaphor onto the genre history of the novel: it works self-flatteringly. Readers of novels, it fondly assumes, became cleverer after Henry James, and so did novelists. The terms of this claim rely on the idea that by the late nineteenth century the novel had – to use the developmental language of such criticism, now mercifully not as fashionable as it was – 'come of age', 'reached maturity', 'emerged', etc. It had learned, that is, to be (more and more successfully) like an implied modern reader. The reader since Henry James has become more skilful at reading Henry James; that is true enough. Unnoticed, though, some other skills have been lost: the skill of writing and reading a novel as a conscious religious exercise in the preparation for death, one function of *Clarissa* for Samuel Richardson's first readers; the skill of communicating and receiving a nimbus of allusive ironies and erotic information through the use of classical allusion, as used by Henry Fielding in, for example, *Joseph Andrews*.

In most transfers of 'evolution' onto other domains 'evolution' is invoked as a justification of human political, social and psychic choices. In this usage, upward development is assured. So-called 'natural selection' in such displacements is really *artificial selection naturalized*. Wilful choices and political urgencies are presented as if they were aspects of an inevitable natural process. For example, when the model of evolution is applied to the whole domain of literature, it can be used to bolster the authority of the current literary canon – and to fix it (against more creative evolutionary principles). Thus, in the recent so-called literary 'wars of the canon' the argument could go as follows: natural selection allows the best to survive and to become normative; the 'great works' we study are those that time has winnowed from the chaff; they survive because they are fit to survive. And I personally have no wish to lose any of them.

The problem comes with the implication, derived from natural selection in its Spencerian articulation, that those works *not* in the current canon are *not* fit to survive, *not* worth reading. Recently, feminist publishing and that of other excluded groups has given the lie to such irreversible patterns. Fortunately, unlike organisms, printed books rarely become extinct, though they may be disregarded. Copies remain on shelves somewhere to be read anew.

Books can also return, meaning something other to a new generation than they meant to the contemporaries who first read them. That is one indication of the intriguing story by Jorge Luis Borges, 'Pierre Menard, Author of Don Quixote' where the work *Don Quixote*, conceived now, becomes a different book from the one written in the seventeenth century, though the words are identical. In the quotation following, Borges's narrator speaks first then alternates with quotations from Menard:

> It is a revelation to compare the *Don Quixote* of Menard with that of Cervantes. The latter, for instance, wrote (*Don Quixote*, Part One, Chapter Nine):
> [. . . *original*]

> [... truth, whose mother is history, who is the rival of time, depository of deeds, witness of the past, example and lesson to the present, and warning to the future.]
>
> Written in the seventeenth century, written by the 'ingenious layman' Cervantes, this enumeration is a mere rhetorical eulogy of history. Menard, on the other hand, writes:
>
> [... *original*]
>
> [... truth, whose mother is history, who is the rival of time, depository of deeds, witness of the past, example and lesson to the present, and warning to the future.]
>
> History, *mother* of truth; the idea is astounding. Menard, a contemporary of William James, does not define history as an investigation of reality, but as its origin. Historical truth, for him, is not what took place; it is what we think took place. The final clauses – *example and lesson to the present, and warning to the future* – are shamelessly pragmatic.
>
> Equally vivid is the contrast of styles.
>
> (*Ibid.*, p. 36)

Imbued with the concerns of a new time (here pragmatism, but just as possibly, feminism 'mother of truth'), Cervantes's words become fraught with twentieth-century reference and emerge, Borges suggests, from a new pen. This imaginary author Menard is writing, *writing out*, with great creative pains, word for word, whole sections of *Don Quixote*. Even copying, Borges suggests, implies expanding and changing, not merely replicating, because the copyist *writes out* in a different era for new readers in a new environment. The whole fable provides an imaginative questioning, parsimonious yet fundamental, of the appropriated evolutionary metaphor which argues that literature 'develops'. Rather, old texts can mean fresh things as fully as current ones.

An example of a much more disastrous displacement of the idea of natural selection, particularly relevant to the work of specific novelists, is the relation of evolutionary theory to race theory. The notion that human society has followed over time an inevitable process of development from the primitive to what nineteenth-century ethnographers called 'modern European man' was taken to carry with it an implicit permission for colonial settlement and for the extinction of the indigenous group. Mid nineteenth-century evolutionary theory could be called upon to reinforce that position. Such thinking, of course, long pre-dated Darwin but, in particular, the reification produced by Herbert Spencer's 'survival of the fittest' seemed to offer scientific authentication of later racist theory. Joseph Conrad gropes through some of its outcomes in the gloom of *Heart of Darkness* where the collusions between Western commerce and Western self-image as 'civilized man' despoil Africa and at the same time cleave rapaciously to its seeming promise of 'primitive' release. Yet, in the *Origin*, at the opening of Chapter 4, 'Natural Selection', Darwin had insisted that 'natural selection' could work most harmoniously when the indigenous inhabitants of a terrain had the time and seclusion to vary and so find multiple ecological niches without interruption or intrusion from outside.

> In such case, every slight modification, which in the course of ages chanced to arise, and which in any way favoured the individuals of any of the species, by better adapting them to their altered conditions, would tend to be preserved; and natural selection would thus have free scope for the work of improvement.
>
> (*Ibid.*, p. 166)

Contrary to what Tim Ingold suggests in Chapter 5 of this volume, Darwin did not always think that the brain capacity of tribes was specific, and limited, to their current position in the world order. Indeed, in his encounters with the Fuegians, first those who were on board *H.M.S. Beagle* after an enforced sojourn of a year and more in London, and then those in the lashing rain and smoke of Tierra del Fuego, he recognized already in the mid 1830s the degree to which human intellectual capacity surpasses the circumstances which control its expression.

The crossed ideas in Darwin's evolutionary theory

The very different aspects of evolutionary theory taken up in other fields can give us a useful map of the discordant concepts at work in 'evolution', both in scientific and cultural settings. Darwin's work is shot through with the pain of powerful crossed ideas. He thinks with three crucial concepts. The three crossed ideas are hyperproductivity, variability and selection. They do not all point in the same direction. Hyperproductivity and variability suggest a copious, generous world reaching out always into unforeseen potential. The abundance of hyperproductivity is necessary if populations are not to wither away; but it contains its own threats of spoliation. Variability launches many diverse individuals into the environment; it is not likeness to the parent type but deviance from it that proves to be the creative principle. Variability does not privilege the normative; the odd, the outrageous, the 'seeming monster' with the power of reproduction are all potentially of greater value than the common form. Set athwart this liberal tumult is the more frugal principle of selection, which insists on aptness to the demands of the immediate environment for survival. Darwin, in his writing, took great pains to distinguish between artificial selection – that is, culturally induced selection for breeding purposes which subdues other creatures to the advantage of humankind – from his newly formulated concept 'natural selection'. Natural selection, he argues, works for the good of each, does not subordinate one species to the will of another and proceeds by slow unconscious processes, not by wilful intervention as does artificial selection. Darwin struggled to divest natural selection of cultural components; that is why he distinguished artificial selection, and also sexual selection.

So my insistence that nineteenth-century 'evolution' (particularly when it trails ideas of development and complexification) is a hampering model to apply to the story of the novel *as a genre* does not at all deny the importance of evolutionary ideas for individual writers of fiction. Fiction is in the nature of a thought experiment, which hypothesizes outcomes without immediate consequences for the reader. As Darwin writes on 28 December 1834 in *The Voyage of the Beagle* (p. 311): 'The limit of man's knowledge in any subject possesses a high interest, which is perhaps increased by its close neighbourhood to the realms of the imagination.'

Many particular novelists have responded to, tested, as well as resisted evolutionary patterns for narrative, and have engaged with the strong ideas that have permeated our culture since Darwin. To cite a very few examples: Thomas Hardy's *Tess of the D'Urbervilles* studies the social decay of a family 'extinct in the male line' and the emergence from it of the fruitful Tess, an 'almost standard woman' the promise of whose human and genetic excellence is spoiled by social violation and prejudice. Conrad's *Victory* and, in a different way, his *Heart of Darkness* probe the potential violent recursiveness always present in those who believe themselves at the apex of civilized progress. James Joyce in *Ulysses* mingles in 'The Oxen of the Sun' the history of language with the history of a child, from conception to birth. These works do not simply mimic Darwinian theories. They may reinscribe them, in a Borgesian sense, copying to arouse new implications. More often they are in contention with them, and are also raising to the surface troubles within the theory itself.

To turn now more directly to the imaginative worlds triggered within and askance Darwin's writing. It is paradoxical, like so many elements in this story, that in ordinary speech now 'evolution' has become a catch-all word which justifies change of any kind. It promises (or seems so to do) that what might seem vacillating is purposeful and developmental (the evolution of the National Health Service, or the United Nations, for example). This promise of foreseeable plan is paradoxical in itself since Darwinian theory promised no such thing. Indeed, he needed to undermine the language of natural theology with its emphasis on 'design' and 'creation' in order to make room for a production-based theory with no set or single outcome. Darwin's work emphasized that the future cannot be known – and that it cannot be known precisely *because of* evolutionary process, with its emphasis on multiplicity, variability and inter-responsiveness of organisms. Evolutionary ideas long precede the nineteenth century. But in the nineteenth century they came to accord with the high value being placed in other areas of social life (such as industrial practices) on *change*. They do not imply, necessarily, controlled sequence. Catastrophic change can also be imagined.

Earlier evolutionary narratives took the single life cycle as a model. Within the life cycle of the organism transformations occur: larvae yield dragon-flies, babies become adults. These transformations are securely sequenced. If the organism does not achieve them it is because it has died on the way. So transformism affirms teleology: a plan is enacted. The German *Bildungsroman* (or development novel), with its emphasis on the growth to adulthood and accommodation with society, describes a similar movement, in which transformations yield coherence, as in Goethe's *Wilhelm Meister*.

But move that process of transformation away from ontogeny to phylogeny (to the development of the species rather than the single organism) and very different and disturbing possibilities emerge. Above all, the future of the species cannot be foreseen. It is not tightly sequenced in an inevitable grammar of development *at species level*.

On The Origin of Species By Means of Natural Selection Or the Preservation of Favoured Races in the Struggle for Life (to give the work its full, perturbing title): the story Darwin tells there has neither beginning nor end. He declares outright that he is not concerned with life's beginning. The immense obliterations of the past give him no encouragement to foresee the future. Indeed, the power to survive for a species is the outcome of innumerable individuals' slight mutations, some of which happen to be in accord with the current demands of the environment and so allow the individual to breed progeny. If those demands remain the same, some of its progeny bearing the same characteristics will in turn be advantaged, or if the demands have altered, quite *others* of the brood will survive. It is all very chancy, highly contingent and not much comfort to the human person – who in this work is never represented in the book's argument, save for one sentence. Other men (all men), however, do work hard in the text as Darwin's multiple informants (editing a new edition of the book recently, I tracked down over 120 informants cited in the text, many of whom were his contemporaries and known personally to Darwin, including 'my son' and 'a man I implicitly trust', probably a servant).

Darwin is concerned with procedures and processes (natural selection, artificial selection, sexual selection). The narrative retrospect of his work mixes time and space freely: not in the past only but in the present world, he argues, different evolutionary activities are occurring. Simple and complex co-exist; not all the drive is towards difficulty or high complexity. The often quoted last paragraph of the *Origin* emphasizes an upward movement ('from so simple a beginning endless forms most beautiful and most wonderful have been, and are being, evolved'). But elsewhere in the work Darwin recognizes that an organism held in a satisfactory and unvarying medium has no need of change. Darwin emphasizes subtle and unconscious shifts. Yet, of course, his language is freighted still with terms that carry the expectation of intervention and plan (selection and preservation are two notorious examples that he struggled with and against in later editions of the work: to say nothing of 'Nature').

The place of the human

In Darwin's writing and in the career of his thought the human presence is ambiguous, at times it seems not even to be there. That is what renders his evolutionary theory so compelling. In Darwinian evolutionary theory the human is not central, yet its predicates remain human. Humankind cannot escape the human condition. We are locked into it by language, which always refers back to human measure. Evolutionary theory has suggested ways, fissured with dread and repulsion as well as pleasure, for human beings to understand themselves in relation to other forms of life. It has named a process, reckless of human self-esteem, by which the human is caught into large and interactive changes. But human beings, on the whole, think most easily in terms of our place, our time, our kin, rather than across all the possible zones of wider interaction with other organisms. The novel is the most *positioned* of literary forms, breeding detail, instance and instantaneity. It is also, as a form, obsessed with the problem of change: a family romance in which progeny's *difference* from parents is of more significance than is likeness to them.

The absence/presence of the human in Darwinian theory as Darwin presents it has been a temptation to many interpreters within and beyond fiction. The apparitional quality of the human in his argument in the *Origin* produces a kind of flirtation that makes readers and commentators desperate to restore humankind to a stable centrality. It is revealing that the most frequent miscitation of the title of Darwin's work in advertisements is as *The Origin of THE Species*: *the* species neatly recentres the human; who else could *the* species be?

Not that that abstention drove out human reference from the text or from its thought: how could it? As Marx wrote to Engels: 'It is remarkable how Darwin recognizes among beasts and plants his English society with its division of labour, competition, opening up on new markets, "inventions", and the Malthusian "struggle for existence".' It is equally revealing, of course, that Marx picked up from the discourse only those elements that chimed in with his own concerns.

Darwin read fiction throughout his life, as his letters and reading notebooks make clear – Jane Austen, for example, before she was fashionable. In his middle and later years he liked to have novels read aloud to him, and he and Emma spent some time most days in this way. Indeed, in the *Descent* he quotes the German philosopher Arthur Schopenhauer to justify an interest in 'love intrigues' as 'really of more importance than all other ends in human life' – because they imply 'the composition of the next generation . . . the weal or woe of the human race still to come is still at stake'. Darwin was, unnecessarily, ashamed of this predilection for novel reading (perhaps the more ashamed because most of the writers he preferred were women) and in his Autobiography he cited it as evidence of affective decline. What he most liked were stories with a happy ending (for that reason he found the weight of compunction in George Eliot hard to bear, though he delighted in *Adam Bede* when it first appeared). The stories *he* had cast into the world with his writing had no such happy ending. Instead, an unplumbed beginning, inexorable process, irreversible events and uncertainty at the end. Or so, looking back in the light of Thomas Hardy's work or George Gissing's, one may cast the narrative grammar implicit in and partly learned from Darwin. But, at the time, Darwin's writings could also generate an insistence on improvement, reach and the resolution of difficulty, including the driving out of those seen as less apt to the modern world –

though even in the work of novelists such as Rider Haggard sadness is admitted as the earlier tribe gives way to the incursions of the colonizer.

Economic theory, language theory, race theory, human encounter with indigenous others, all entered the crucible of Darwin's thinking. 'Wherever the European has trod, death seems to pursue the aboriginal' he wrote in the *Voyage* of his sojourn in New Zealand. Indeed, I would argue, his experience of revolution, uprising, genocidal wars, colonization and its effects, all play into the making of the concept of natural selection as thoroughly as did his encounters with the birds and turtles of the Galapagos Islands. So, too, did his reading of certain novelists, particularly Walter Scott, whose novels explore the processes of change by which one nation-culture gives way to the incoming forces of another.

What Scott demonstrates is that not military force only but subtle and prolonged shifts of expectation, pressure of economic powers, land possession, may undermine the energies of an indigenous group. He showed these processes at work – and showed, too, the stamina with which they continue to be resisted – in novels such as *Waverley* and *Old Mortality*. Scott's novels tend, with infinite reluctance, to naturalize the process of cultural extinction. Darwin recognized a pattern that seemed parallel to this when he visited van Diemen's land in the *Beagle*. That pattern of understanding, mingling compunction and a sense of inevitability, is available in Scott – and permits colonization to proceed. But during the years that he was preparing the *Origin* Darwin read too, and avidly, the works of Harriet Martineau, friend of his brother Erasmus; in particular he perused her historical novel *The Hour and the Man*, whose hero is Toussaint-L'Ouverture, leader of the slave uprising in the Indies. Martineau's sympathies are racked across the slaves' right to freedom and the brutality of revolution, but the book is focused on the great figure of the insurgent Toussaint. His revolution fails but it is seen as a phase in the inevitable rise of his people.

So when we encounter evolutionary theory claimed as a model for intellectual, aesthetic or even advertising ventures we need to recognize that Darwinian evolutionary theory itself was not an autonomous 'scientific' discovery, achieved by controlled empirical methods alone. Evolutionary thinking is not a grid; it is a bundle of apprehensions. Coming before genetics, the major initiating formulations in Darwin's work drew upon multiple resources, by no means all of them scientific. The strength of his theorizing, indeed, is in the eclecticism, even the bricolage, of Darwin's habits of mind. In his youth everything

109

interested him; and as he grew older, particulars never ceased to fascinate him and yield the unexpected observation. A mind as ranging and capacious as this is bound to encounter contradictions; perhaps it may also absorb them rather too easily. Total coherence according to the tenets of Baconian enquiry is not Darwin's way. Certainly, the contradictory implications of Darwinian theory have led to its being read out in multiple directions and to confusion at many crossroads.

Darwin's evolutionary theory is a system disturbed at once by empathy and by obliteration (Darwin speaks of the infinite lost generations of the past, the improbability of any current species persisting far into the future). That has been one of its principle resources for writers of fiction who seek to heave lost persons, possibilities, creatures, out of oblivion, as Hardy marks the shine on wood in the dairy produced by the forgotten rumps of cows easing themselves against it over generations. Or there is the mummers' play in *The Return of the Native*, where some of the experienced participants dramatize it because it is a ritual thoroughly known. Memory is the stuff of fiction; the lost generations, past thought, are the necessary ballast of Darwin's theories too. Empathy (*Eingefühl*) is feeling from within. As Darwin described plants, animals, birds and insects he entered not the individual life so much as the interactions between lives. Relations fascinated and perturbed him: 'Relation of organism to organism the most important of all relations' he gave as a summary heading in the *Origin*. He thought in classes, aware of the gaps as well as the affinities between species, and within species. He thought also with the activity of tendrils, alert to the delicate nudging of growth, the responsiveness of reach. His eye noted always the exception. His theory was concerned with generation and descent, the transformation of populations. The medium of that descent, however, was in his thinking always of the individual organisms. Those organisms are transient, lost to mind, occasionally stamped as fossils, but more often surviving on the earth only as loam or dung or dust.

Abundance and hyperproductivity obliterate the process of loss, thickening the present so that the unthinkableness of past life itself vanishes from consciousness. All this Darwin tells us in the *Origin*. What he does not articulate, since reflexivity is not his concern, is that abundance and hyperproductivity are at once material to his theory and comforting. His awareness of loss is at once evidential and emotional: it is *inconvenient* that so little survives to demonstrate his contentions (soft matter perishes); it is intimately *distressing* that oblivion is so profound. He knew, or thought he knew, how little might survive in memory from his own experience of his mother's death when he was eight: he later found that he had only a few, frail, impersonal memories of her. The present *needs* to be full.

The death of the human

Darwin's theory at the same time demanded death, the death of large numbers, *and* the medium of the individual. In that conflict lay his challenge also, to novelists such as Theodore Dreiser beset with new urban populations set down in the expanses of America.

Books can become part of the experience of many who have not even read them if the stories they sustain tell sufficiently deeply on the desires and fears of succeeding societies. That, I would argue, happened to Darwin's writing. Borges in *Other Inquisitions* used evolutionary metaphors to describe this process in relation to the work of H. G. Wells:

> The Time Machine, The Island of Dr. Moreau, The Plattner Story, The First Men in the Moon ... are the first books I read, perhaps they will be the last. I think that they will be incorporated, like the fables of Theseues or Ahasuerus, into the general memory of the species and even transcend the fame of their creator or the extinction of the language in which they were written.
>
> (*Other Inquisitions*, p. 88)

Of the several points of contestation produced by – and extended out from – Darwinian theory the one that most fascinated Wells was the doubtful identification of development with the idea of progress. Wells was a pupil of the scientist Thomas Huxley and, through Huxley, he saw that evolution carried no absolute promise of improvement – a position that Huxley himself reached late in his career when he foresaw the 'universal winter' where 'only such low and simple organisms as the Diatom of the arctic ... and the Protococcus of the red snow' will survive.

Darwin was irresolute on this point, longing to believe in improvement and asserting it in the psalmody of the *Origin*'s last paragraph. Yet, in his Autobiography he acknowledged how his hopes were chilled, not only by pressures within evolutionary thinking, but by the rise alongside it of another powerful set of ideas that seemed to countermand all efflorescence towards perfection:

> It is an intolerable thought that he [*man*] and all other sentient beings are
> doomed to complete annihilation after such long-continued slow progress.
>
> (*Autobiographies of Charles Darwin and T. H. Huxley*, pp. 153–4)

As novelists moved in upon the knot of evolutionary ideas towards the end
of the century they could work with, and against, the contradictory implications
of Darwinian thinking: is this a theory that scotches teleology or one that refig-
ures the idea of plan? Is it a story that has no particular place for humankind
or one that draws on human activities for its figuring of all relations? Is it a
story of progress or decay? Beyond Darwin's own work came its merging with
Spencerian sociology and the eugenicist insistence on the 'survival of the fittest'
applied to human life. But that tussle of contradictions within the theory and
among its accretions was not all. The pressures were further intensified and
complicated by the apparent contradictions *between* evolutionary theory and
the new physics of the time.

Like Jack London and Edgar Rice Burroughs later, Wells was fascinated by
what Huxley had called in his 1863 anthropological essays *Man's Place in
Nature*. In *The Time Machine*, published in 1896, his traveller enters a period of
the future when the human species has divided into two: the effete and graceful
Eloi, with their childlike insouciance, who live in pleasure above ground, and
the oppressed and dangerous Morlocks who live half-blinded and whitened
among the underground machines that keep this civilization blithe. The work
brilliantly moves through a series of interpretations by the narrator, each of
them displaced by a more threatening and sardonic knowledge as he proceeds.
At first he is simply disappointed by the lack of intellectual progress he finds:

> You see I had always anticipated that the people of the year Eight Hundred
> and Two Thousand off would be incredibly in front of us in knowledge, art,
> everything. Then one of them suddenly asked me a question that showed him to
> be on the intellectual level of one of our five-year-old children.
>
> (*Ibid.*, p. 25)

Then, he perceives the analogy with the danger of revolution from the
oppressed: the 'graceful children of the Upper World were not the sole descendants of our generation, but that this bleached, obscene, nocturnal Thing . . . was
also heir to all the ages' (*ibid.*, p. 47). At last, he perceives a balance of
oppression and terror between the two species. The graceful Eloi are meat for

the Morlocks, seized under cover of dark and consumed beneath ground. Are they, indeed, not aesthetes but a kind of cattle? Both species are bound into this locked balance. Neither has versatility, the mark of the human and of evolutionary potential: 'It is a law of nature we overlook, that intellectual versatility is the compensation for change, danger, and trouble. An animal perfectly in harmony with its environment is a perfect mechanism. Nature never appeals to intelligence until habit and instinct are useless.' There is 'no intelligence where there is no change' (*ibid.*, pp. 78–9). In the last section of the book that changelessness takes a yet more extreme form, one that draws directly on Huxley's recognition that evolution need not imply progress, and on Darwin's aghast awareness that energy rather than gathering more and more potential is running down.

> So I travelled, stopping ever and again, in great strides of a thousand years or more, drawn on by the mystery of the earth's fate, watching with a strange fascination the sun grow larger and duller in the westward sky, and the life of the old earth ebb away
>
> (*Ibid.*, p. 85)

Here, before the discovery of radioactivity, is the typical late Victorian dread of the death of the sun and the acting out of the second law of thermodynamics. *That* is the key contradiction for the period: between the evolutionary expectation of a more refined and energetic future, honed by natural selection, set against the loss of available energy through entropy until the earth reaches equilibration, the great quiet that is universal death.

Primate novels

For novelists of the turn of the century the question of the human species and its proximity to other primates becomes the focus also for pleasurable anxiety. Jack London, for example, in *Before Adam* has a first-person narrator who throughout his childhood half resides in the mid Pleistocene Period, sharing with his primate parents there the dominant emotion of fear. The suggestion is that the child has tapped into 'race-memories' and when he gets to college he has an explanation of his dreams:

> But at college I discovered evolution and psychology, and learned the explanation of various strange mental states and experiences. For instance there was the falling-through-space dream – the commonest dream experience, one practically known, by first-hand experience, to all men.

> This, my professor told me, was a racial memory. It dated back to our remote ancestors who lived in trees.
>
> *(Ibid.,* pp. 20–1)

The topic of specification can function as a cover for examining questions of social class and race or ethnicity too. One compelling example is the ever-popular *Tarzan of the Apes*, set in 1888 but published in 1917. In Burrough's story a young aristocratic couple die in the jungle and their infant is adopted by apes. He grows up among them as Tarzan, leader of them all (rather as Mowgli does in Rudyard Kipling's 1894 *Jungle Book*; both books imply a colonizing subtext). At the end of the book Tarzan's aristocratic rather than simian lineage is authenticated. This is achieved by a paradox that only the reader can enjoy. We share his gentlemanly reticence. Tarzan, in love with Jane and she with him, encounters her fiancé. The young man asks him about his family:

> 'If it's any of my business, how the devil did you ever get into that bally jungle?'
> 'I was born there,' said Tarzan quietly. 'My mother was an Ape, and of course she didn't tell me much about it. I never knew who my father was.'
>
> *(Tarzan of the Apes,* p. 269)

A neat final turn: by claiming his ape foster-parents as his blood lineage Tarzan demonstrates to us his actual aristocratic chivalry. Jane marries her fiancé. Tarzan, with gentlemanly self-abnegation, bows out.

Darwinian themes and their contradictory indications, as well as their troubled relations with other strong theory, have not vanished from fiction. Hilary Mantel's *A Change of Climate* (1994) weaves evolutionary issues into family romance: the epigraph she chooses, from *The Descent of Man*, voices both Darwin's claim to neutrality and the anxiety the novelist and the scientist share concerning their responsibility for description, perhaps invention: 'We are not here concerned with hopes and fears, only with the truth as far as our reason allows us to discover it. I have given the evidence to the best of my ability . . .'. She follows that passage with a further epigraph, from Job (iv, 7), which turns back to challenge the first: 'Consider, what innocent ever perished, or where have the righteous been destroyed?' That utterance is then itself challenged by the book which, with gruelling courage demonstrates, like Darwin, that it is necessary to truth-telling to show that the innocent do perish and the righteous are destroyed.

In a teasingly brilliant and mournful novel called *Monkey's Uncle*, also published in 1994, Jenny Diski sets evolutionary ideas in abrasion with chaos theory – neither of them set out in ways that would satisfy a technician bent on exposition but rather, using the model at once of *Alice* and of biography, exploring how each of us is traceried with the ideas that have functioned to compose our common culture.

As her protagonist, Charlotte FitzRoy, descends into madness she finds herself alongside Captain FitzRoy of the *Beagle*, sharing his despair – a despair which the current reader will recognize has been generated out of an example (the infinite coastline) offered in the mathematician Benoit Mandelbrot's *The Fractal Geometry of Nature* (1982). FitzRoy's professional task is to survey the coasts and shores; to live he needs to believe, as a devout Christian, in a controlling pattern for his task and his existence. But as he approaches the shore 'the apparent simplicity altered'. Smooth lines become 'increasingly complex and jagged':

> FitzRoy longed with all his heart for simplicities, but his mind was committed to precision. And *precisely*, the line he might have reproduced from simple looking turned out not to exist in any continuous way, but became manifold . . .
>
> The obvious concavity of a cove became a multitude of ins and outs, set within the greater simpler shape, as coming closer, the outlines of the many rock formations of which it was made up were defined. Then, moving closer still into shallower water, a myriad of individual pebbles on the shoreline grew visible and undulated away all remaining hope of smooth simplicity. Finally, wading on to land, his heart pounding, desperation thudding in his head, FitzRoy sank on his knees on the beach, to the alarm of his men, examining the shingle and even the very grains of sand, each minute one of which also, of course, had an outline of its own – and so many, too many ever to hope to give an account of the pattern they made.
>
> (*Monkey's Uncle*, p. 22)

Though FitzRoy is disturbed by Darwin and repudiates his theories, it is this silent and isolated insight that lies at the root of his despair: multiplicity of pattern and of scale producing, for him, patternlessness. The cast of Diski's knowing fiction include Jenny, the orang-utan Darwin visited at the London Zoo, the Fuegians Jemmy Button and Fuegia Basket being returned to their land by FitzRoy and Darwin, and the three old gentlemen of the nineteenth century, Marx, Freud and Darwin, condemned to comfort themselves through eternity with frequent picnics and irascible conversation. Tribute and satire, the novel is a funeral offering of the liveliest kind to the lost determinisms of the three overlapping and competing thought-systems into which recent European generations have been born. The comic melancholy of Diski's tale composes an elegy for those secure reductionisms.

But, as I hope I have demonstrated, Darwin's writing always shakes loose from reductionism. 'People often talk of the wonderful event of intellectual man appearing – the appearance of insects with other senses is more wonderful.' So wrote the young Darwin in his notebook, fascinated as ever by diversity, otherness and the world without (or outwith) the human. What to others seems supplement to him is centre. He leaves always something further to think with and against – something for novelists to try out language against, as they insist, repeatedly, rightly, on bringing the human back over the brink of the horizon.

FURTHER READING

Beer, G. (ed.) *Autobiographies of Charles Darwin and T. H. Huxley*, Oxford: Oxford University Press, 1974.

Beer, G. 'The death of the sun: Victorian solar physics and the solar myth', in *The Sun is God: Painting, Literature and Mythology in the Nineteenth Century*, ed. B. Bullen, pp. 159–80, Oxford: Oxford University Press, 1989.

Borges, J. L. 'The first Wells', in *Other Inquisitions 1937–1952*, pp. 86–8, London: Souvenir Press, 1973.

Borges, J. L. 'Pierre Menard, Author of Don Quixote', in *Ficciones*, with an introduction by John Sturrock, pp. 29–38, London: Everyman, 1993.

Browne, J. and Neve, M. (ed.) Darwin, C. *The Voyage of the Beagle*, by Charles Darwin (first published 1839), London, 1989.

Burckhardt, F. and Smith, S. (eds.) *Correspondence of Charles Darwin*, Cambridge: Cambridge University Press, 1988.

Burroughs, E. R. *Tarzan of the Apes*, London: Macmillan, 1917.

Campbell Ross I. (ed.) *The Life and Opinions of Tristram Shandy*, by L. Sterne, vol. V (first published 1761), Oxford: Oxford University Press, 1983.

Darwin, C. *The Descent of Man and Selection in Relation to Sex*, London, 1871.

Diski, J. *Monkey's Uncle*, London: Weidenfeld & Nicolson, 1994.

Huxley, T. H. 'The struggle for existence in human society', in *Evolution and Ethics* (essay first published 1888), pp. 195–236, London: Macmillan, 1906.

Keymer, T. *Richardson's Clarissa and the Eighteenth Century Reader*, Cambridge: Cambridge University Press, 1992.

London, J. *Before Adam*, London, n.d.

Moorcock, M. (ed.) *The Time Machine*, by H. G. Wells (first published 1896), London, 1994.

Peckham, M. (ed.) *The Origin of Species by Charles Darwin: A Variorum Text*, Philadelphia, 1959.

Ryazanskaya, S. (ed.) *Marx–Engels Selected Correspondence*, Moscow, 1965.

7   The Evolution of Science

FREEMAN DYSON

Analogies

I was asked to write about the 'Evolution of Science'. This is an enormous subject and would take a historian to do it justice. I am not a historian. I am a scientist with a smattering of knowledge about history. I prefer to write about things I know. Here, I tell stories rather than digging deep into the sources of historical truth. I write about astronomy, which is one little corner of science, and about recent events with which I am familiar. I use the recent history of astronomy to illustrate some evolutionary themes, which may or may not be valid when extended to earlier periods or to other areas of science.

My approach to evolution is based on analogies between biology, astronomy and history. I begin with biology. The chief agents of biological evolution are speciation and symbiosis. In the world of biology these words have a familiar meaning. Life has evolved by a process of successive refinement and subdivision of form and function; that is to say, by speciation, punctuated by a process of bringing together alien and genetically distant species into a single organism, i.e. symbiosis. As a result of the work of the biologist Lynn Margulis and other pioneers, the formerly heretical view, that symbiosis has been the mechanism for major steps in the evolution of life, has now become orthodox. When we view the evolution of life with an ecological rather than an anatomical perspective, the importance of symbiosis relative to speciation becomes even greater.

As a physical scientist, I am struck by the fact that the borrowing of concepts from biology into astronomy is valid on two levels. One can see in the sky many analogies between astronomical and biological processes, as I shall shortly demonstrate. And one can see similar analogies between intellectual and biological processes in the evolution and taxonomy of scientific disciplines. The evolution of the universe and the evolution of science can be described in the same language as the evolution of life.

Speciation in the sky

In the context of astronomy, speciation occurs by the process of phase transition. A phase transition is an abrupt change in the physical or chemical properties of matter, usually caused by heating or cooling. Familiar examples of phase transitions are the freezing of water, the magnetization of iron, the precipitation of snow from water vapour dissolved in air. In many of these, the warmer phase is a uniform disordered mixture while the cooler phase divides itself into two separate components with a more ordered structure. Such transitions are called order–disorder transitions, and humid air changing to cold dry air plus snowflakes is a typical example. Snowflakes are a new species, with a complex crystalline structure that was absent from the humid air out of which they arose. Also, by the action of the earth's gravity, snowflakes spontaneously separate themselves from air and fall to the ground. At all stages in the evolution of the universe we see order–disorder transitions with the same two characteristic features: first, the sudden appearance of structures that did not exist before; and, second, the physical separation of newborn structures into different regions of space.

Another name for the process of phase transition from disorder to order is symmetry-breaking. From a mathematical point of view, a disordered phase has a higher degree of symmetry than an ordered phase. For example, the environment of a molecule of water in humid air is the same in all directions, while the environment of the same molecule after it is precipitated into a snowflake is a regular crystal with crystalline axes oriented along particular directions. The molecule sees its environment change from the greater symmetry of a sphere to the lesser symmetry of a hexagonal prism. The change in the environment from disorder to order is associated with a loss of symmetry. Sudden loss of symmetry is seen in many of the most important phase transitions as the universe evolves.

In the earliest stages of its history, the universe was hot and dense and rapidly expanding. Matter and radiation were then totally disordered and uniformly mixed. One of the greatest symmetry-breakings was the separation of the universe into two phases: one contained most of the matter and was destined to condense later into galaxies and stars; the other contained most of the radiation and was destined to become the intergalactic void. The separation happened as soon as the universe became transparent enough, so that large lumps of matter pulled together by their own gravitation could radiate away their gravitational energy into the surrounding void. As a result of this transition, the universe lost its original spatial symmetry. Before the transition, it had the symmetry of uniform space. After the transition, it became a collection of irregular lumps. The same process of symmetry-breaking was then repeated successively on smaller and smaller scales. A single lump of the first generation was a huge mass of gas, locally uniform and locally symmetrical. The local uniformity of the gas was then broken when it condensed into the second-generation lumps which we call galaxies. The gas in a local region of a galaxy cooled further until it condensed into the third-generation lumps which we call giant molecular clouds. Finally, the gas and dust in a local region of a molecular cloud condensed into the fourth-generation lumps which we call stars and planets. The universe in this way became a hierarchical assortment of lumps of various shapes and sizes. The formation of lumps was at each stage driven by gravity and assisted by phase transitions allowing the physical separation of matter into different phases.

The processes of astronomical speciation did not stop after the stars and planets were formed. After the earth had condensed out of the interstellar dust, a new world of opportunities opened for separation of phases and growth of structures. First came the separation of the interior of the earth into its main components: core, mantle and crust. Next came the separation of the earth's surface into land, ocean and atmosphere. This is a continuing process, with water constantly circulating from the ocean into the atmosphere, onto the land and back to the ocean. The third process transforming the earth is the division of the crust into plates and the formation and destruction of the crust at the plate boundaries, the process known as plate tectonics; plate tectonics is a powerful force, constantly giving the earth new structures. The fourth process creating structure and order on earth is the most powerful of all. The fourth process is life. Life appeared here between three and four billion years ago and gave the concept of speciation a new meaning.

The transition from dead to living was a phase transition of a new type. It was a transition from disorder to order, in which the ordered phase acquired the ability to perpetuate itself after the conditions that caused it to appear had changed. There are many theories of the origin of life, and there is no direct evidence to decide which theory is true. All that we know for sure is that a complicated mixture of organic chemicals made the transition to an ordered phase that could grow and reproduce itself and feed on its surroundings. And then, after the ordered phase was once established, it possessed the flexibility to mutate and evolve into a million different species. Life has given to our planet a richness of structure that we see nowhere else in the universe. But the diversification of new forms of life on the earth is in many respects similar to the diversification of new celestial species, galaxies and dust clouds, and stars and planets, in the universe as it was before life appeared. The evolution of life fits logically into the evolution of the universe. Both in the non-living universe and on the living earth, evolution alternates between long periods of metastability and short periods of rapid change. During the periods of rapid change, old structures become unstable and divide into new structures. During the periods of metastability, the new structures are consolidated and fine-tuned while the environment to which they are adapted seems eternal. Then the environment crosses some threshold that plunges the existing structures into a new instability, and the cycle of speciation starts again.

Symbiosis

Phase transitions are one of the two driving forces of evolution. The other is symbiosis. Symbiosis is the reattachment of two structures, after they have been detached from each other and have evolved along separate paths for a long time, so as to form a combined structure with behaviour not seen in the separate components. Symbiosis played a fundamental role in the evolution of eukaryotic cells from prokaryotes. The mitochondria and chloroplasts that are essential components of modern cells were once independent free-living creatures. They first invaded the ancestral eukaryotic cell from the outside and then became adapted to living inside. The symbiotic cell acquired a complexity of structure and function that neither component could have evolved separately. In this way symbiosis allows evolution to proceed in giant steps. A symbiotic creature can jump from simple to complicated structures much more rapidly than a creature evolving by the normal processes of mutation and speciation.

Symbiosis is as prevalent in the sky as it is in biology. Astronomers are accustomed to talking about symbiotic stars. The basic reason why symbiosis is important in astronomy is the double mode of action of gravitational forces. When gravity acts upon a uniform distribution of matter occupying a large volume of space, the first effect of gravity is to concentrate the matter into lumps separated by voids. The separated lumps differentiate and evolve

separately. They become distinct species. Then, after a period of separate existence, gravity acts in a second way to bring lumps together and bind them into
pairs. The binding into pairs is a sporadic process depending on chance
encounters. It usually takes a long time for two lumps to be bound into a pair.
But the universe has plenty of time. After a few billion years, a large fraction
of objects of all sizes become bound in symbiotic systems, either in pairs or in
clusters. Once they are bound together by gravity, dissipative processes bring
them closer together. As they come closer together, they interact with one
another more strongly and the effects of symbiosis become more striking.

Examples of astronomical symbiosis are to be seen wherever one looks in the
sky. On the largest scale, symbiotic pairs and clusters of galaxies are common.
When galaxies come into contact, their internal evolution is often profoundly
modified. A common sign of symbiotic activity is an active galactic nucleus. An
active nucleus is seen in the sky as an intensely bright source of light at the
centre of a galaxy. The probable cause of the intense light is gas falling into a
black hole at the centre of one galaxy as a result of gravitational perturbations
by another galaxy. It happens frequently that big galaxies swallow small
galaxies. Nuclei of swallowed galaxies are observed inside the swallower, like
mouse-bones in the stomach of a snake. This form of symbiosis is known as
galactic cannibalism.

On the scale of stars, we can distinguish many types of symbiosis, because
there are many types of star and many stages of evolution for each of the stars
in a symbiotic pair. The most conspicuous symbiotic pairs have one component
that is highly condensed (a white dwarf, a neutron star or a black hole) and the
other component a normal star. If the two stars are orbiting around each other
at a small distance, gas spills over from the normal star into the deep gravitational field of the condensed star. The gas falling into the deep gravitational
well becomes intensely hot and produces a variety of unusual effects, recurrent
nova outbursts, intense bursts of X-rays and rapidly flickering light variations.
The more common and less spectacular symbiotic pairs consist of normal stars
orbiting around each other close enough so that the mass is exchanged between
them.

The rarest type of symbiotic pair consists of two condensed stars. These can
be seen with radiotelescopes if one component of the pair is a pulsar, a neutron
star emitting radio pulses as it rotates. One such pair, a symbiosis of two neutron stars, was discovered by the radio-astronomers Joseph Taylor and Russell

Hulse who received the Nobel Prize for Physics in 1993 for the discovery. This symbiotic pair of neutron stars is scientifically important because it gave us the first clear evidence for the existence of gravitational waves. The drag produced by gravitational waves brings them steadily closer together as time goes on. Ultimately they will be brought so close together that they become dynamically unstable and fall together into a single star with a splash of spiral arms carrying away their angular momentum. The process of collapse takes only a few thousandths of a second and must result in a huge burst of outgoing radiation. The details of the collapse have been calculated by Fred Rasio, a young astronomer now at the Massachusetts Institute of Technology. The collapse of symbiotic neutron stars may explain the mysterious bursts of gamma-rays that are seen coming from random directions in the sky at a rate of about one per day. I will have more to say later about the way gamma-ray bursts were discovered. If Fred Rasio's explanation of the gamma-ray bursts is correct, they are the most violent events in the whole universe, even more violent than the supernova explosions that occur when neutron stars are born. A symbiotic pair of neutron stars can deliver a stronger punch than any single star by itself. Symbiosis becomes more and more central as the universe evolves.

From our human point of view, the most important example of astronomical symbiosis is the symbiosis of the earth and the sun. The system of sun and planets and satellites is a typical example of astronomical symbiosis. At the beginning, when the solar system was formed, the sun and the earth were born with different chemical compositions and physical properties. The sun was made mainly of hydrogen and helium; the earth was made of heavier elements. The sun was physically simple, a sphere of gas heated by the burning of hydrogen and shining steadily for billions of years. The earth was physically complicated, partly liquid and partly solid, its surface frequently transformed by phase transitions. The symbiosis of these two contrasting worlds made life possible. The earth provided chemical and environmental diversity for life to explore. The sun provided physical stability, a steady input of energy on which life could rely. The combination of the earth's variability with the sun's constancy provided the conditions in which life could evolve and prosper.

Tools and concepts

I now move from astronomy to history, from the evolution of the universe to the evolution of science. The major events in the history of science are called scientific revolutions, and of these there are two kinds – those driven by new concepts and those driven by new tools. They are analogous to biological revolutions driven by speciation and by symbiosis, or to astronomical revolutions driven by phase transition and by gravitational binding. When a field of science is overturned by a new concept, the revolution starts from the inside, from an internal inconsistency or contradiction within the science, and results in a phase transition to a new way of thinking. When a field of science is overturned by new tools, the revolution starts from the outside, from tools imported from another discipline, and results in a symbiosis of the two disciplines. In both types of revolution, the final outcome is a new subdiscipline of science and a new species of scientist, specialized in the new ideas or in the new tools as the case may be.

Thomas Kuhn, in his famous book *The Structure of Scientific Revolutions* (1962), talked almost exclusively about concepts and hardly at all about tools. His idea of a scientific revolution is based on a single example, the revolution in theoretical physics that occurred in the 1920s with the advent of quantum mechanics. This was a prime example of a concept-driven revolution. Kuhn's book was so brilliantly written that it became an instant classic. It misled a whole generation of students and historians of science into believing that all scientific revolutions are concept driven. The concept-driven revolutions are the ones that attract the most attention and have the greatest impact on the public awareness of science, but in fact they are comparatively rare. In the last 500 years we have had five major concept-driven revolutions, associated with the names of Copernicus, Newton, Darwin, Einstein and Freud, besides the quantum-mechanical revolution that Kuhn took as his model. During the same period there have been about twenty tool-driven revolutions, not so impressive to the general public but of equal importance to the progress of science. I will not attempt to make a complete list of tool-driven revolutions. Two prime examples are the Galilean revolution resulting from the use of the telescope in astronomy, and the Watson–Crick revolution resulting from the use of X-ray diffraction to determine the structure of big molecules in biology. Galileo brought into astronomy tools borrowed from the emerging technology of eyeglasses. James Watson and Francis Crick brought into biology tools borrowed from physics. The effect of a concept-driven revolution is to explain old things in new ways. The effect of a tool-driven revolution is to discover new things that have to be explained. In astronomy there has been a preponderance of tool-driven revolutions. We have been more successful in discovering new things than in explaining old ones.

Up to this point I have been discussing generalities; now I turn to the details, as I happen to be more interested in the details of particular scientific revolutions than in the general rules that they may or may not exemplify. The details are real. The general rules are at best an approximation to reality, at worst a delusion. Several tool-driven astronomical revolutions happened in the nineteenth century. One was the introduction of high-resolution spectroscopy by Joseph von Fraunhofer, allowing astronomers to study the chemical composition of the sun and the stars. Another was the development of astronomical photography by Henry Draper and James Keeler, allowing astronomers to study with long exposures objects a thousand times fainter than the human eye could see. In each case, the old community of sky-watchers absorbed by symbiosis an alien technology with different traditions. Fraunhofer belonged to the world of commercial glass manufacture. Photography brought into the observatories experts trained in the craft of studio portraiture. The symbiosis of sky-watchers with these two alien cultures resulted in the emergence of a new science with the name 'astrophysics', the science that tries to describe quantitatively the physical processes going on in stars and other celestial bodies. I pass briefly over the nineteenth century because I want to have some time left for the twentieth.

Bernhard Schmidt and Fritz Zwicky

Let us examine in detail three twentieth-century revolutions. The first is associated with the names of Bernhard Schmidt and Fritz Zwicky. Schmidt invented a new kind of telescope and Zwicky understood how to use it. Schmidt and Zwicky were both highly unorthodox characters. Schmidt was an optical technician who grew up on a small island in the Baltic, experimented as a boy with home-made explosives, blew off his right hand at the age of twelve and then taught himself the art of making telescopes with his left hand. He supported himself by selling mirrors of superlative quality to amateur astronomers and professional observatories all over Europe. In 1932 he built at Hamburg the first telescope of the new type now known simply as a 'Schmidt'. The Schmidt was a revolutionary instrument. By throwing overboard the customary way of designing optical systems, Schmidt obtained images in sharp focus over a field of view a hundred times larger than the field of view of conventional telescopes.

This meant that it was possible for the first time to produce sharp photographs of large areas of sky quickly and conveniently. For the first time it was possible to scan the entire sky photographically in a reasonable time and at reasonable cost. Schmidt was a man of few words. His collected works fill three pages.

Fritz Zwicky was a young Swiss physicist, working at the California Institute of Technology, when Schmidt invented his telescope. Zwicky was interested in supernovae, the new stars that occasionally shine in the sky for a few weeks with extraordinary brilliance. Until that time very few supernovae had been identified. One had been seen by Tycho Brahe and another by Kepler, before the days of telescopes, but it was not clearly established that they were different from ordinary novae. Zwicky was one of the few people who took supernovae seriously. He understood, before this became accepted dogma, that supernovae were cataclysmic events on a totally different scale from ordinary novae. He understood that a supernova was an event of extreme violence, probably resulting in the disruption of an entire star. He saw that the key to the understanding of supernovae was to observe a substantial number of them rather than one or two. And he saw that the Schmidt telescope was the tool he needed, the tool that would make it possible to find supernovae in reasonable numbers and to study them systematically.

About thirty years later, Zwicky wrote an autobiography with the title *Discovery, Invention, Research through the Morphological Approach*. He believed passionately in a private theory of everything, a theory that he called the morphological method. The idea of the morphological method is that you write down a complete list of all the conceivable ways of solving a problem before you choose the way that actually solves the problem. If you judge the method by the number of important things that Zwicky discovered, you have to conclude that it is highly effective. The disadvantage of the method is that it does not seem to work so well if your name is not Zwicky.

Here is Zwicky's description, recorded in his autobiography, of how he used the morphological approach to study supernovae.

> I wish to caution all hotheads that it is not advisable to try to do everything at the same time, a mistake which is often committed by individuals and by institutions whose funds are limited. For instance, the construction of multipurpose telescopes is in general not to be recommended. It is better to concentrate one's attention on specific problems and to build instruments best adapted for their solution. One often discovers subsequently that such instruments can also be used effectively for other purposes. As an example I mention

the eighteen-inch Schmidt telescope on Palomar Mountain, whose construction I promoted in 1935 for the specific task of supernovae . . . I put this instrument into operation on the night of September 5, 1936, and immediately started a systematic survey of several thousand galaxies.

(*Ibid.*, p. 91)

As soon as Zwicky heard about Schmidt's invention, he moved fast, with the enthusiastic help of George Hale, to acquire an 18-inch Schmidt and install it in a small dome on the Palomar site that was to be the future home of the 200-inch telescope. Zwicky's little Schmidt was the first telescope on Palomar Mountain. It was the first Schmidt telescope to be installed anywhere in the world in a place with clear skies and good seeing. It is still there today, and still doing important science. Zwicky had it all to himself, a situation that he regarded as essential for doing serious work in astronomy. He had a single assistant who was paid to work for him full-time. He ran a programme that became a prototype for all the later sky surveys carried out with bigger instruments and bigger budgets. He understood that in order to detect rare and transient events it was necessary to scan the whole sky repeatedly, over and over again. For five years he and his assistant Johnson took pictures of huge areas of sky, night after night, covering the northern sky as often as they could. They compiled a catalogue of 50 000 galaxies and 10 000 clusters of galaxies that they kept under observation. About once every three months, comparing an image of one of these galaxies with an earlier image of the same galaxy, they found a newly bright feature that they could identify as a supernova. Supernova candidates were then studied in detail and their spectra analysed with bigger telescopes. Working in this way for five years, between 1936 and 1941, Zwicky and Johnson discovered twenty supernovae. On the basis of this sample, Zwicky could measure roughly the frequency of occurrence of supernovae in the universe and their absolute optical brightness. He identified the two main types of supernova. Supernovae were moved suddenly from the shadowy edge of astronomy to the well-observed centre.

The Schmidt–Zwicky revolution had consequences extending far beyond the initial discoveries. It caused a major shift in our perception of the universe as a whole. The old Aristotelian view of the celestial sphere as a place of perfect peace and harmony had survived intact the intellectual revolutions associated with the names of Copernicus, Newton and Einstein. The Aristotelian view still dominated the practice of astronomy until 1935. Zwicky was the first astronomer who imagined a violent universe. He chose to study supernovae because they provided the most direct evidence of violent processes occurring on a universal scale. After 1935, the idea of a universe dominated by violent events gradually spread, until it was confirmed by the spectacular discoveries of radio-astronomers and X-ray astronomers thirty years later. Now we all take it for granted that we live in a violent universe. But this awareness only began in 1935 with the little Schmidt telescope on Palomar Mountain.

Vela Hotel

Twenty years after the Schmidt–Zwicky revolution came two more tool-driven revolutions, two symbiotic invasions of traditional astronomy by borrowed technologies, first by radiotelescopes and then by X-ray telescopes. I skip over the radio-astronomy and X-ray revolutions, because their history is well known and I have nothing new to say about them. I consider instead another revolution, the gamma-ray revolution, which led thirty years later to the launching of the Compton Gamma-Ray Observatory now orbiting over our heads.

A few years before the gamma-ray revolution, Zwicky had stated his rule: do not build multipurpose telescopes but concentrate your attention on specific problems and build instruments best adapted for their solution. Zwicky said it would often happen that a single-purpose instrument would afterwards find other unexpected applications. The gamma-ray revolution was a fine example confirming Zwicky's rule. It began in the Los Alamos National Laboratory with a project called Vela Hotel, designed to verify compliance with the 1963 Limited Test Ban Treaty. Vela Hotel deployed satellites in orbits far beyond geosynchronous, carrying among other things gamma-ray detectors that would respond sensitively to nuclear explosions in space or in the upper regions of the earth's atmosphere. The gamma-ray detectors never detected any bomb tests. As Zwicky had surmised, they turned out to be ideally suited to detect natural events of an unexpected kind. They detected bursts of gamma-rays arriving mysteriously from unknown sources, unconnected with any human activity or with any known astronomical object. As a result, a part of the weapons-dominated culture of Los Alamos was symbiotically absorbed into the peaceful culture of astronomy.

The first discovery of gamma-ray bursts was published in 1973 by the Los Alamos physicists R. W. Klebesadel, I. B. Strong and R. A. Olson. After describing the Vela Hotel detectors, they say, 'This capability provides continuous

coverage in time which, combined with isotropic response, is unique in observational astronomy.' A proud claim and a true one. It was true that no astronomical instrument up to that time had ever been capable of detecting signals
for twenty-four hours a day over the entire sky. The Vela Hotel detectors had
three additional advantages that distinguished them from earlier instruments.
They recorded events with high time-resolution, with accurate absolute timing,
and with four independent detectors at points widely separated in space. As a
consequence of the good timing and wide separation, most of the observed
events could be located on the sky with reasonable precision. The superior
capabilities of the Vela Hotel instruments arose naturally out of the require-
ments of the nuclear weapons culture. The astronomical culture, before Vela
Hotel, had never seen a need for such capabilities.

I have vivid memories of a visit to Los Alamos when Ian Strong talked to me
about the early evidence for gamma-ray bursts. This was after the first Vela
Hotel discoveries but before the first publication. Strong was reluctant to publish, not because the data were secret but because they seemed too weird to be
credible. The Los Alamos team delayed their publication for four years after
the first bursts were seen. The four-year delay is a measure of how revolutionary the discovery of gamma-ray bursts was felt to be. The discoverers thought
their data would be more credible if they could identify a few of the gamma-ray
burst sources with unusual objects visible in optical or radio wavelengths. In
spite of strenuous efforts, they failed for ten years to find convincing identifi-
cations. As usually happens when a new window into the universe is opened,
the view was so strange that it took considerable courage to publish it.

It was a happy accident that the Vela Hotel satellites combined so many features that were well matched to the gamma-ray burst phenomenon. They had
high orbits, continuous all-sky sensitivity and multiple detectors widely separated in space. Unfortunately, because of the constraints imposed by the use of
the Space Shuttle as a launch vehicle, the Compton Gamma-Ray Observatory
was forced to abandon every one of these advantages. It has a low orbit, with
almost half of its view of the sky obscured by the earth, and no ability to measure direction of a source by differential timing at the ends of a long base-line.
It was designed, in violation of Zwicky's rule, as a general-purpose observatory.
We may hope that future generations of gamma-ray burst detectors will be spe-
cial-purpose instruments and will fully exploit the advantages of the Vela Hotel
architecture. The Vela Hotel revolution will not be complete until it is extended

to the study of transient events in other parts of the electromagnetic spectrum besides gamma-rays. We should use the Vela Hotel architecture to search for transient events with detectors of visible light, infra-red, ultra-violet and X-rays, deployed on multiple small satellites in high orbits with long base-lines. And the satellites should be complemented with ground-based detectors searching for transient events in other channels, such as radiowaves, neutrinos and gravitational waves. The Vela Hotel revolution still has a long way to go.

Digital astronomy

The next revolution after Vela Hotel is the digital astronomy revolution. It belongs to the present rather than to the past. We are living in the midst of it. It is driven by another new tool of observation, the charge-coupled device, popularly known as the CCD. This revolution was predicted by Fritz Zwicky, long before the CCD was invented. I quote from the Halley Lecture given by Zwicky in 1948 at Oxford University, with the title 'Morphological Astronomy' (pp. 126–7). To save space I have omitted some phrases and sentences, but I have not added a single word.

> The photo-electronic telescope introduces the following new features. (1) Electrons can be accelerated from the image surface to the recording surface and power can be fed into the telescope to increase the intensity of the signals . . . (2) Uniform background of light . . . may be eliminated by electric compensation . . . The sky background . . . may thus be scanned away . . . (3) Although the original image may move, dance or scintillate . . . because of the unsteadiness of the atmosphere, the refocused image on the recording surface can . . . be steadied . . . Zworykin has actually built such an image stabilizer . . . (4) Automatic guiding of a telescope may be accomplished . . . (5) Images from photo-electronic telescopes can be televised, and the search for novae, supernovae, variable stars, comets, meteors, etc., can be put on a mass production scale.

Zwicky was hoping in 1948 that all these good things could be achieved by a television camera system that he had been working on with his friend Vladimir Zworykin at the Radio Corporation of America (RCA). Zworykin was my nextdoor neighbour in Princeton, a great engineer and a cantankerous character, almost as eccentric as Zwicky. The RCA camera did not fulfil Zwicky's hopes. Now the CCD does everything that he wanted. The main reason why the RCA system failed was that it still depended on photographic plates for recording images. The main reason why the CCD succeeds is that it is coupled to a digital memory instead of to a chemical image on a plate. The digital astronomy revolution had to wait until the technology of image-processing had matured, with powerful microprocessors and digital memories to match the abundance of data that the CCD could supply.

The digital astronomy revolution is now in full swing. Astronomy is now an intimate symbiosis of three cultures, the old culture of optical telescopes, the newer culture of electronics and the newest culture of software engineering. One of the results of this symbiosis is the Sloan Digital Sky Survey (SDSS), a project in which many of my colleagues at Princeton are actively engaged. The SDSS is a modern version of the Palomar Sky Survey, the photographic survey of the northern sky which was finished in 1956 and supplied the astronomers of the world with their first accurate large-scale map of the universe. The Palomar Sky Survey plates have been enormously useful but are now about to be superseded by something better. The output of the SDSS will be a photometrically precise map of the sky in five colours, plus a collection of spectra providing red-shifts of about a million galaxies and other interesting objects. One by-product of this output will be a three-dimensional view of the large-scale structure of the universe over a volume 100 times as large as the volume covered by existing surveys. Another by-product will be a catalogue of about 100 000 quasars, gravitational lenses, brown dwarfs and other peculiar objects, giving a complete count of objects in each category down to some faint limiting magnitude. The entire output of the survey will be transmitted at electronic speed to any astronomical centre possessing a digital memory large enough to swallow it. The size of memory required will be measured in tens of terabytes, a terabyte being a million megabytes. For customers lacking such a gargantuan memory, various predigested versions of the output will be provided, with the photometric data compressed into star catalogues and galaxy catalogues supplemented by images of particularly interesting local areas. The essential difference between the SDSS and all previous surveys is that the output will be linear, consisting of directly measured light intensities instead of measured marks on a photographic plate. The output will be packaged so that all the tricks of modern data-processing can be immediately applied to it.

The Sloan Digital Sky Survey is a collaborative project in which Princeton is one of seven partners. It uses a new 2.5-metre wide-field telescope, built in New Mexico and dedicated to the project for five years. With luck, the survey will be finished by the year 2002. A large array of CCD detectors sits in the focal plane of the telescope. The hardware components of the project do not stretch the state of the art in telescope or detector design. The main novelty of the project lies in the software, which has to control the sequence of operations, calibrate the CCD detectors, monitor the sky quality and apply several levels of data-compression to the output before distributing it to the users. The major share of the cost of the project is paid by the Sloan Foundation, following the good example of the National Geographical Society, which funded the Palomar Sky Survey fifty years earlier. The total cost is estimated to be $50 million, including the capital cost of the telescope. This is about a half of the cost of a major ground-based observatory, and about a thirtieth of the cost of the Hubble Space Telescope.

After our little Digital Sky Survey is finished, there will be other surveys putting into digital memory larger and deeper maps of the universe. There are many directions for future surveys to explore. One survey may push towards fainter and more distant objects, another towards higher angular resolution, another into a wider choice of wavelengths, another into higher spectral resolution. The power and speed of digital data-processing will continue to increase. The digital astronomy revolution will continue to give us clearer and more extended views of the large-scale structure of the universe. There will be no natural limit to the growth of digital surveys, until every photon coming down from the sky is separately processed and its precise direction and wavelength and polarization recorded.

Finally, I want to touch on space science. Here, even more than in ground-based astronomy, the digital revolution has created enormous opportunities which have not been fully exploited. Space missions on a grand scale, such as the Voyager explorations of the outer planets and the Hubble Space Telescope explorations of distant galaxies, have sent back to earth a wealth of scientific knowledge. But the cost of such missions is out of proportion to their scientific value. From a purely scientific point of view, neither Voyager nor Hubble was cost-effective. Both missions were launched in a political climate which valued them as symbols of nationalistic glory rather than as scientific tools. Now the winds of political change are blowing hard. Space scientists are keenly aware that times are changing. Billion-dollar missions are no longer in style. Funding in the future will be chancy. The best chances of flying will go to missions that are small and cheap.

In 1995 I spent some weeks at the Jet Propulsion Laboratory (JPL) in California. JPL built and operated the Voyager missions. It is the most independent and the most imaginative part of NASA. I was particularly interested in two proposals for planetary missions that JPL wished to fly, the Pluto Fast Fly-by and the Kuiper Express. Both missions existed as ideas in the minds of JPL designers. The Pluto Fast Fly-by would complete the Voyager exploration of the outer planets by taking high-resolution pictures of Pluto and its satellite Charon. The Kuiper Express would similarly explore the Kuiper Belt of newly discovered planetary objects orbiting the sun beyond the orbit of Pluto. Both missions are based on a radical shrinkage of the instruments that were carried by Voyager. The digital revolution has made radical shrinkage possible. I held in my hands the prototype package of instruments for the new missions. The package weighs seven kilograms. It does the same job as the Voyager instruments, which weighed half a ton. All the hardware components, optical, mechanical, structural and electronic, have been drastically reduced in size and weight without sacrifice of performance.

Daniel Goldin, the Administrator of NASA, encouraged JPL to design these new missions, to carry on the exploration of the outer solar system with spacecraft radically cheaper than Voyager. Each Voyager mission cost about a billion dollars. The JPL designers came back to Goldin with their design for the Pluto Fast Fly-by. Their estimated cost for the mission was 700 million dollars. According to hearsay, Goldin said, 'Sorry, but that is not what I had in mind'. The mission was not approved. The Pluto Fast Fly-by missed its chance for a quick start. It failed because it did not depart radically enough from the design of Voyager. It still carried for its electrical power supply the heavy Voyager thermo-electric generator using the radioactivity of plutonium-238 as the source of energy. It still relied on massive chemical rockets to give it speed for the long haul from here to Pluto. It was new wine in an old bottle, new instruments riding on an old propulsion system. The instruments were radically shrunk, but the rest of the spacecraft was not shrunk in proportion.

Meanwhile, a new design with the name Pluto Express has been cobbled together by combining pieces of the old Pluto Fast Fly-by and Kuiper Express. The Pluto Express is new wine in a new bottle. It is the first radically new planetary spacecraft since the early Pioneers went to Venus. The Pluto Express uses solar-electric propulsion to give it high speed. The propellant is xenon, which can be conveniently carried as a supercritical liquid as dense as water without refrigeration. The prototype xenon-ion engine was undergoing endurance tests in a tank at JPL when I visited. It must run reliably for eighteen months without loss of performance before it can be seriously considered for an operational mission. The power source for the mission is a pair of large and extremely light solar panels. The panels are large enough to provide power for instruments and for communication with earth as far away from the sun as the Kuiper Belt. No plutonium generator is needed. The Pluto Express has finally jettisoned the last heavy piece of Voyager hardware, so that it can fly fast and free.

The Pluto Express is a daring venture, breaking new ground in many directions. It demands new technology and a new style of management. It may fail, like the Pluto Fast Fly-by, because its designers make too many compromises. Its designers may not dare enough. But solar-electric propulsion has opened the door to a new generation of cost-effective small spacecraft, taking full advantage of the digital revolution. If the Pluto Express fails to fly, some other more daring mission will succeed. The use of solar-electric propulsion will change the nature and style of planetary missions. Spacecraft using solar-electric propulsion may wander around the solar system, changing their trajectories from time to time to follow the changing needs of science. Solar-electric propulsion will make them adaptable as well as small and cheap. The new generation of spacecraft will evolve from Voyager as birds evolved from dinosaurs. In space science, just as in evolutionary biology or in international politics, the collapse of the old order opens new opportunities for adventurous spirits.

FURTHER READING

Klebesadel, R. W., Strong, I. B. and Olson, R. A. 'Observations of gamma-ray bursts of cosmic origin', *Astrophysical Journal Letters* **182** (1973), L85. [This letter announced the discovery of gamma-ray bursts, whose nature is still one of the major mysteries of astronomy.]

Kuhn, T. *The Structure of Scientific Revolutions*, Chicago: Chicago University Press, 1962. Second edition, 1970.

Margulis, L. *Symbiosis in Cell Evolution*, San Francisco: Freeman and Co., 1981. [This is the classic statement of the case for symbiosis as a major driving force of evolution.]

Margulis, L. and Dolan, M. F. 'Swimming against the current', *The Sciences*, January–February (1997), 20–5. [This short essay describes some dramatic new evidence that emerged after Margulis's book was published, further extending the evolutionary role of symbiosis.]

Zwicky, F. *Discovery, Invention, Research through the Morphological Approach*, Toronto: Macmillan, 1969. [This was originally published in German

(Muenchen-Zuerich, Droemersche Verlagsanstalt, 1966) – a mixture of autobiography and science, displaying Zwicky's eccentricity as well as his brilliance.]

Zwicky, F. 'Morphological astronomy', *The Observatory* **68** (1948), 121–43. [This was Zwicky's Halley Lecture, delivered at Oxford on 12 May 1948. It is a brief and less contentious statement of his scientific philosophy, full of remarkable insights.]

8 The Evolution of the Universe

MARTIN REES

Cosmologists study evolution on the grandest scale of all. They aim to set our earth and our solar system in an evolutionary scheme stretching right back to the formation of the Milky Way galaxy – right back even to a so-called 'Big Bang' that set our entire observable universe expanding and imprinted the physical laws that govern it.

Evolution within our galaxy

Let us start with something fairly well understood – the life cycle of our sun, a typical star. About 4.5 billion years ago it condensed from an interstellar cloud, and contracted until the centre became hot enough to ignite fusion of hydrogen into helium. This process will keep it shining until, after another five billion years, the hydrogen runs out. The sun will then flare up, becoming large enough to engulf the inner planets, and to vaporize all life on earth. After this 'red giant' phase the inner regions contract into a white dwarf – a dense star no larger than the earth, though nearly a million times more massive.

We are quite confident about these calculations because the relevant physics has been well studied in the laboratory – atomic and nuclear physics, Newtonian gravity and so forth. Astrophysicists can just as easily compute the life cycles of stars with half the sun's mass, or twice, four times, etc. Heavier stars burn brighter, and trace out their life cycle more quickly.

Stars live so long compared to astronomers that we are granted just a single 'snapshot' of each star's life. But we can test our theories, by looking at the whole *population* of stars. Trees can live for hundreds of years. But for a newly landed Martian who had never seen a tree before, it would take no more than an afternoon wandering around in a forest to deduce the life cycle of trees: from looking at saplings, fully grown specimens and some which had died.

In the Orion Nebula, for instance, new stars are even now condensing within glowing gas clouds. The best 'test beds' for checking such calculations are globular clusters – swarms of a million different stars, held together by their mutual gravity, which all formed at the same time.

But not everything in the cosmos happens slowly; sometimes stars explode catastrophically as supernovae. The closest supernova of the twentieth century occurred in 1987. Its sudden brightening and gradual fading have been followed not only by optical astronomers (Figure 1) but by those using the other modern techniques – radio, X-ray and gamma-ray telescopes – which have opened new 'windows' on the universe.

In about 1000 years, it will look like the Crab Nebula (Figure 2), the relic of a supernova witnessed and recorded by Chinese astronomers in A.D. 1054. Now, nearly a thousand years later, we see the expanding debris from the explosion. The Crab Nebula will remain visible, gradually expanding and fading, for a few thousand years; it will then become so diffuse that it merges with the very dilute gas and dust that pervades interstellar space.

Cosmic alchemy

Supernovae fascinate astronomers, but why should anyone else care about explosions thousands of light years away? Because, were it not for supernovae, there would be no planets, still less any complex evolution on them.

Of the ninety-two chemical elements that occur naturally, some are vastly more common than others. For every ten atoms of carbon, you would find, on average, twenty of oxygen, and about five each of nitrogen and iron. But gold is 100 million times rarer than oxygen, and others – uranium, for instance – are rarer still. Why are carbon and oxygen common, but gold and uranium so rare? This everyday question is not unanswerable – but the answer involves ancient stars that exploded in our Milky Way more than five billion years ago, before our solar sytem formed.

Stars much heavier than the sun evolve in a more complicated and dramatic way. After they have used up their central hydrogen (and turned into helium) gravity squeezes them further. Their centres get still hotter, until helium atoms can themselves stick together to make the nuclei of heavier atoms: carbon (six protons), oxygen (eight protons) and iron (26 protons). A kind of 'onion skin' structure develops: where the hotter inner layers have been transmuted further up the periodic table.

When their fuel has all been consumed (when their hot centres are

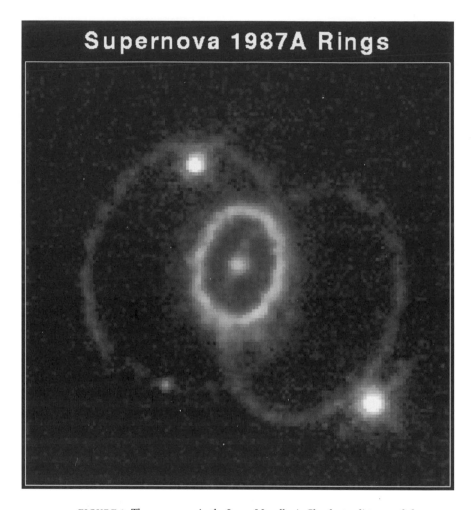

Supernova 1987A Rings

FIGURE 1 The supernova in the Large Magellanic Cloud, at a distance of about 160 000 light years. This picture, taken with the Hubble Space Telescope about five years after the actual explosion was observed, shows strange 'rings', which are thought to result from interaction between the radiation and debris from the explosion and external material that was probably ejected as a slower wind before the supernova occurred.

transmuted into iron) big stars face a crisis. A catastrophic infall squeezes their centres to the density of an atomic nucleus, triggering an explosion that blows off the outer layers. This explosion manifests itself as a supernova of the kind that created the Crab Nebula. The debris contains the outcome of all the

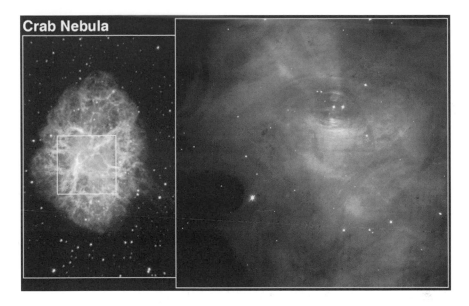

Crab Nebula

FIGURE 2 The Crab Nebula. The right-hand frame shows continuing activity in the central part of the nebula, induced by the spinning neutron star (pulsar) that was left behind after the explosion.

nuclear alchemy that kept the star shining over its entire lifetime – a great deal of oxygen and carbon, plus traces of many other elements. The calculated 'mix' is gratifyingly close to the proportions now observed in our solar system.

The Milky Way, our home galaxy, resembles a vast ecosystem. Pristine hydrogen is transmuted, inside stars, into the basic building blocks of life: carbon, oxygen, iron and the rest. Some of this material returns to interstellar space, thereafter to be recycled into new generations of stars.

A carbon atom, forged in an early supernova, might wander for hundreds of millions of years in interstellar space. It might then have found itself in a dense interstellar cloud, which collapsed under its own gravity to form stars. It may have entered the core of some new massive star, where it is processed further up the periodic table (into silicon, or into iron), and then flung out in another supernova. Or it may have joined one of the less massive stars, each surrounded by a spinning gaseous disc that condenses into a retinue of planets. One such star could have been our sun. The same carbon atom may have found itself in the newly forming earth, perhaps eventually in a human cell. Each atom has a pedigree extending back far earlier than our solar system's birth. We are literally the ashes of long-dead stars.

But how did our galaxy itself emerge? Where did the basic hydrogen come from? To answer these questions we must broaden our horizons still further in both space and time to the universe of galaxies.

The visible universe

Galaxies are held in equilibrium by a balance between two effects: gravity, which tends to make the stars all gather together; and the countervailing effect of the stellar motions, which if gravity did not act would make a galaxy fly apart. In some galaxies, our own and Andromeda among them, 100 billion stars move in nearly circular orbits in discs. In others, the less photogenic ellipticals, stars are swarming around in more random directions, each feeling the gravitional pull of all the others. Galaxies are not as well understood as stars. Indeed, as I will explain, we do not even know of what they are primarily made.

Galaxies interest cosmologists because they are 'test particles' for probing structure and motions in the large-scale universe. The nearest few thousand galaxies – those closer than about 300 million light years – have been mapped in both hemispheres. They are irregularly distributed into clusters and super-clusters. Are there, you may ask, clusters of clusters of clusters *ad infinitum*? Our universe is not like that. If it were, we would see conspicuous clumps in the sky, however deep into space we probed. Although the nearest few thousand galaxies are conspicuously clumped, the brightest *million* galaxies are actually fairly uniform over the sky; as we look at still fainter galaxies, probing still greater distances, clustering becomes less evident and the sky appears smoother.

There is, in other words, a well-defined sense in which the universe is broadly homogeneous. A terrestrial analogy may clarify this. The ocean displays complex patterns – waves (sometimes small riding on large), foam, etc. But once your gaze extends beyond the scale of the longest ocean swells, you see an overall uniformity, stretching to the horizon many miles away. A patch of ocean large enough to be 'typical' must obviously extend several times further than the scale of the longest waves. Our horizon extends far enough to encompass many patches statistically similar one to another, each large enough to constitute a 'fair sample'.

This broad-brush uniformity of *seascapes* is not, however, a general feature of *landscapes*: on land, progressively larger mountain peaks may stretch all the way to the horizon, and a single topographical feature may dominate the entire view. Cosmology is, by definition, the study of the entire universe. We can see only one universe – probably, indeed, only a tiny part of everything there is. Despite these limitations, scientific cosmology *has* progressed, but only because our observable universe (the volume out to the 'horizon' of our observation) resembles a seascape rather than a mountain landscape. Even the biggest superclusters are still small in comparison with the range of powerful telescopes.

The overall motions in our universe are simple too, as Edwin Hubble first realized. Distant galaxies recede from us with a speed proportional to their distance, as though they all started off packed together ten to fifteen billion years ago.

Far away towards the horizon, we see domains whose light set out when the universe was more compressed, more closely packed together (Figure 3). Astronomers can actually *see* the remote past. Telescope images reveal huge numbers of very faint galaxies each so far away that its light set out before our solar system formed. Even more remote are the *quasars* – hyperactive centres of a special class of galaxies, so bright that they vastly outshine the 100 billion stars in their host galaxy. The 'distance record' is held by a quasar so red-shifted that the Lyman-alpha 1216 Å line, the strongest feature in the spectrum of hydrogen and in the far ultra-violet, reaches us in the red part of the spectrum, at around 7200 Å. The ratio of the observed to the emitted wavelength, 5.89, is the factor by which the universe has expanded since the light set out.

Evidence for a Big Bang

Quasars are probes of the era when galaxies were young, and perhaps just forming. But what about still earlier epochs? Did everything really start with a so-called 'Big Bang'? The idea goes back to the Belgian Catholic priest Georges Lemaître in 1931. The phrase itself was introduced by Fred Hoyle, as a derisive description of a theory he never liked.

The name has stuck, and the clinching evidence for the theory came in 1965, when Arno Penzias and Robert Wilson found excess microwave noise, coming equally from all directions and with no obvious source, in their antenna at Bell Laboratories in New York. This has momentous implications: intergalactic space is not completely cold; it is about 3 K above absolute zero (0 K, −273 °C).

141

FIGURE 3 This picture shows a very deep exposure taken with the Hubble Space Telescope. Although it images only a small patch of sky – about a thousandth the area covered by the full moon – it reveals hundreds of very faint objects. Most of these are galaxies so far away that their light set out when they had only recently formed.

That may not seem much, but there are about a billion quanta of radiation (photons) for every atom in the universe.

This 'cosmic background' causes some 1% of the background 'fuzz' on a television set. It is an 'afterglow' of a pregalactic era when the entire universe was hot and dense and opaque. After the universe had expanded for about half a million years, the temperature fell below 3000 K; the primordial radiation then shifted into the infra-red. The universe literally entered a dark age, which persisted until the first stars in the first galaxies, and maybe also the first quasars, formed and lit up space again. The expansion cooled and diluted the radiation, and stretched its wavelength. But it would still be around – it fills the universe and has nowhere else to go!

But we have got firm grounds for believing that the temperature was once billions of degrees, not just thousands – hot enough for nuclear reactions. The rapid expansion did not allow enough time for everything to be processed into iron, as in hot stars. However, about 25% would turn into helium. The rest would still be hydrogen apart from traces of deuterium and lithium.

What is remarkable is that the proportion of helium in old stars and nebulae, now pinned down with 1% accuracy, turns out to be just about what has been calculated. As a bonus, so are the proportions of lithium and deuterium as well. Moreover, these particular elements were a problem for the stellar nucleogenesis scenario that was so successful for carbon, oxygen, etc. This corroborates an extrapolation right back to when the universe was hot enough for nuclear reactions to occur, i.e. when it was just a *few seconds old*.

One day in 1992, what was then in my opinion the best British daily paper (*The Independent*) heralded a cosmological discovery with this front page (Figure 4); it is a comprehensive depiction of cosmic evolution. (Even dinosaurs are featured; I suppose they are the only scientific topic that matches cosmology in popular appeal.) It goes right back to 10^{-43} second. This is what is called the Planck time, when everything was so dense that quantum fluctuations were important for the entire universe.

So, can you believe all the cosmology reported in the newspapers? Is the universe indeed evolving as depicted here? Over the last few years, the case for a 'Big Bang' has had several boosts: the COBE (Cosmic Background Explorer) satellite showed that the background radiation had the expected spectrum, to a precision of a part in 10 000, and there have been better measurements of cosmic helium and deuterium. Moreover, there are several discoveries that *might* have been made, which would have invalidated the hypothesis, and which have *not* been made. The Big Bang has lived dangerously for twenty-five years, and survived.

The grounds for extrapolating back to the stage when the universe had been expanding for *a second* (when the helium formed) deserve to be taken as seriously as, for instance, ideas about the early history of our earth, which are based

FIGURE 4 The front page of *The Independent* when the detection by the COBE satellite of angular fluctuations in the background radiation was announced. These fluctuations, probably formed in an ultra-early 'inflationary' phase of cosmic expansion, were the precursors of large-scale structure in our present universe.

on inferences by geologists and palaeontologists which are equally indirect (and less quantitative). There are some fervent believers. The great Soviet cosmologist Yakov Zeldovich once claimed that the Big Bang was 'as certain as that the Earth goes round the Sun' (even though he must have known the dictum of his compatriot physicist Lev Landau that cosmologists are 'often in error but never in doubt').

I would bet at least 90% on the general concept – not quite 100%. Consistency does not guarantee truth. Our satisfaction may be as illusory as that of a Ptolomaic astronomer who has just fitted a new epicycle.

Is it absurdly presumptuous to claim to know *anything* about the beginnings of our entire observable universe? Not necessarily. It is complexity, and not sheer size, that makes things hard to understand. In the primordial fireball everything must have been broken down into its simplest constituents. The early universe really could be less baffling – more within our grasp – than the smallest living organism. It is biologists and the Darwinians who face the toughest challenge!

I will return to our universe's hot dense beginnings – and what happened in the first second (the lower part of Figure 4) – but let us now look forward rather than backward, as forecasters rather than fossil hunters.

Futurology

Cosmic time spans extend at least as far into the future as into the past. Suppose America had existed for ever, and you were walking across it, starting on the east coast when the earth formed, and ending up in California ten billion years later, when the sun is about to die. To make this journey, you would have to take *one step every 2000 years*. All recorded history would be three or four steps, just before the half-way stage – somewhere in Kansas perhaps. Not the culmination of the journey!

In this perspective, we are still near the beginning of the evolutionary process. The progression towards diversity has much further to go. Even if life is now unique to the earth, there is time for it to spread from here throughout the entire galaxy, and even beyond.

In about five billion years the sun will die; and the earth with it. At about the same time (give or take a billion years) the Andromeda Galaxy, already falling towards us, will crash into our own Milky Way, merging to form a single amorphous elliptical galaxy. But will the universe expand *forever*, attaining some asymptotic heat death? Or will it, after an immense time, recollapse to the Big Crunch?

The ultra-long-range forecast depends on how much the cosmic expansion is decelerating. The deceleration comes about because everything in the universe exerts a gravitational pull on everything else. It is straightforward to calculate that the expansion will eventually go into reverse if the average cosmic density exceeds about three atoms per cubic metre. Space seems even emptier than that: if the atoms in all the stars and gas in all the galaxies were dispersed uniformly, they would fall short of this 'critical' density by a factor of at least 50.

145

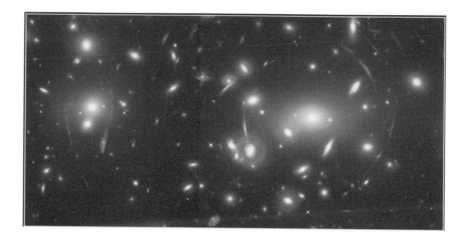

FIGURE 5 The cluster of galaxies Abell 2218. The brighter galaxies belong to the cluster. The faint 'arcs' are remote galaxies, lying far beyond the cluster, whose images are being distorted and magnified by gravitational lensing (a phenomenon whereby light rays are deflected and focused by the gravitational field of a large mass along the line of sight). The strength of the lensing implies that the cluster contains about ten times as much mass in 'dark matter' as in the galaxies we actually see.

At first sight this seems to imply perpetual expansion, by a wide margin. But the case is not so straightforward because there seems to be at least ten times as much material in 'dark' form as we see directly. One line of evidence comes from the discs of galaxies such as our Milky Way or Andromeda. These contain neutral hydrogen gas, which does not itself weigh much, but serves as a tracer of the orbital motion. Radio-astronomers can detect this gas via its emission of the famous 21 cm spectral line. It extends far beyond the limit of the optically detectable disc. The orbital speed is roughly the same all the way out. If the outermost clouds were feeling just the gravitational pull of what we can see, their speeds should fall off roughly as the square root of distance outside the optical limits of the galaxy: the outer gas would move slower, just as Neptune and Pluto orbit the sun more slowly than the earth does. The surprisingly high speed of the outlying gas tells us that an extended invisible halo surrounds these galaxies – just as, if Pluto were moving as fast as the earth, we would have to infer a heavy invisible shell outside the earth's orbit but inside Pluto's.

There is also dark matter pervading entire clusters of galaxies. Figure 5 shows a cluster of galaxies. The faint streaks and arcs are remote galaxies, several times further away than the cluster itself, whose images are, as it were, viewed through a distorting lens. Just as a regular pattern on background wallpaper looks distorted when viewed through a curved sheet of glass, the gravity of the cluster of galaxies deflects the light rays passing through it. This picture would have fascinated Fritz Zwicky, the far-sighted eccentric whom Freeman Dyson extols in Chapter 7. It was Zwicky who first realized, in the 1930s, that clusters of galaxies would fly apart unless they contained more gravitating stuff than is visible; he was also the first person to suggest that gravitational lensing might actually be observable. The visible galaxies in the cluster contain only a tenth as much material as is needed to produce these distorted images – evidence that clusters of, as well as individual, galaxies contain ten times as much mass as we see.

What could this dark matter be? Maybe it is faint stars whose centres are not squeezed hot enough to ignite their nuclear fuel; or black holes, remnants of big stars that were bright when the galaxy was young but have now died. But there are other quite different options. The hot early universe may have contained not just atoms and radiation, but other particles as well. In particular, there should be huge numbers of *neutrinos* – about a billion for every atom in the universe. So even a very tiny individual mass would make the cumulative gravitational effects of neutrinos important. But do neutrinos have any mass at all? Recent experiments at Los Alamos made such a claim but they remain controversial. The results were announced in a paper with thirty-nine authors, but the fortieth member of the group published, in the same issue of the same journal, a paper with a contrary interpretation! So we would be prudent to suspend judgement. If the claimed mass is right, neutrinos contribute more gravitating stuff than do all the stars and gas in the universe.

At least we know neutrinos exist. But particle theorists have a long shopping list of particles that *might* exist, and (if so) could have survived from the early phases of the Big Bang. If such particles pervade our galaxy, there would be 100 000 of them in every cubic metre, most passing straight through the earth without interacting. But their cross-section for colliding with ordinary atoms, though tiny, is not quite zero, and sensitive experiments are being set up to detect the rare events when this happens. The equipment must be placed deep underground, to reduce other types of background signal. A UK group is building such an experiment down a mine in Yorkshire. It is a difficult experiment, but a positive result would not only reveal what 90% of the universe is made of, but also discover new types of particle that could never be detected in other ways.

We should not be surprised that there is dark matter. There is no reason why everything in the universe should shine. The challenge is to decide among many candidates. Its dominance may demote our cosmic status still further. Copernicus dethroned the earth from a central position. Hubble showed that the sun was not in a special place. Now particle chauvinism may have to go. We ourselves, and all the stars and galaxies, would then be trace constituents of a universe whose large-scale structure is controlled by the gravity of dark matter of a quite different kind – we see, as it were, just the white foam on the wavecrests, not the massive waves themselves.

The reliably inferred dark matter in galaxies and clusters is not quite enough to bring cosmic expansion to an eventual halt. However, there is widespread theoretical prejudice (which I will return to later) that the universe has almost exactly the critical density (maybe marginally higher, so that its space–time extent is finite rather than infinite). To those of us who share this prejudice, the burden of proof is on those who contend that there cannot be much still more elusive dark matter *between* clusters of galaxies.

A word now about the emergence of galaxies and clusters. People often wonder how the universe can have started off in thermal equilibrium, a hot dense fireball, and ended up manifestly far from equilibrium: temperatures now range from blazing surfaces of stars (and their even hotter centres) to the night sky only three degrees above absolute zero. Although this seems contrary to thermodynamic intuitions that temperatures tend to equilibrate as things evolve, it is actually a natural outcome of cosmic expansion, and the workings of gravity.

Gravity has the peculiar tendency to drive things further from equilibrium. When gravitating systems lose energy they get *hotter*. A star that loses energy and deflates ends up with a *hotter* centre than before. (To establish a new and more compact equilibrium where pressure can balance a (now stronger) gravitational force, the central temperature must *rise*.)

Gravity does something else. It renders the expanding universe unstable to the growth of structure, in the sense that even very slight initial irregularities would evolve into conspicuous density contrasts. Theorists are now carrying out increasingly elaborate computer simulations of how this happened. Slight fluctuations are 'fed in' at the start of the simulation: exactly how they are prescribed depends on the cosmological assumptions. Figure 6 shows three 'frames' from a simulation of a region containing a few thousand galaxies, large enough to be a fair sample of our universe. As the expansion proceeds, regions slightly denser than average lag further and further behind. Eventually they stop expanding and condense into gaseous protogalaxies which fragment into stars. The same process on larger scales leads to clusters and superclusters. The aim is to make different assumptions about the initial fluctuations, the dark matter, etc., and see which leads to a pattern of structure closest to a typical sample of the real universe.

The microwave background, a relic of a pregalactic universe that was not perfectly smooth, should bear the imprint of the initial fluctuations. This radiation in effect comes from a very distant surface or horizon, far beyond the quasars; it has propagated freely since a time long before the clusters had fully formed. Radiation from an incipient cluster on that surface would appear slightly cooler, because it loses extra energy climbing out of the gravitational pull of an overdense region. Conversely, radiation from the direction of an incipient void would be slightly hotter. The fractional differences in the temperature involve this same small ratio – they are only about one part in 100 000.

Temperature non-uniformities with about this amplitude were first detected by NASA's COBE satellite. To measure such small effects was a technical triumph. But the fluctuations were not unexpected. It would have been far more baffling if they had not been there. That would have implied an early universe so smooth that it would not have been compatible with the conspicuous clustering we see in the present universe – we would have had to postulate some process more efficient than gravity for pulling these structures together.

COBE got the first positive results, but already these are being complemented and extended by ground-based and balloon experiments and two ambitious new space experiments are now planned. The embryonic precursors of galaxies and larger cosmic structures are no longer just hypothetical entities but can actually be measured.

If one had to summarize, in just one sentence, 'What's been happening since the Big Bang?' the best answer might be to take a deep breath and say 'Ever since the beginning, gravity's "anti-thermodynamic" effects have been amplify-

(a)

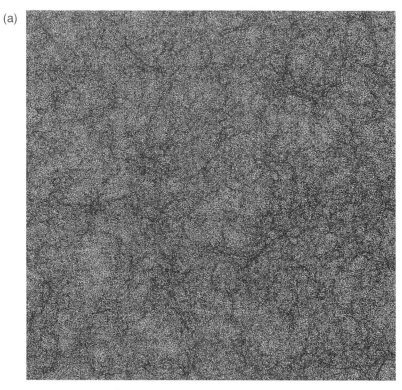

(b)

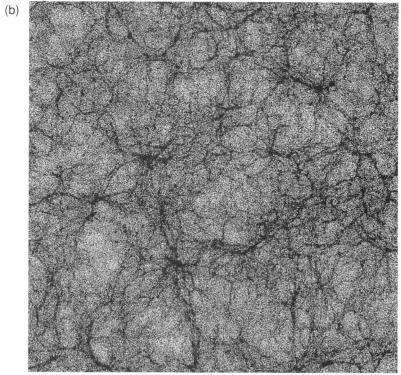

(c)

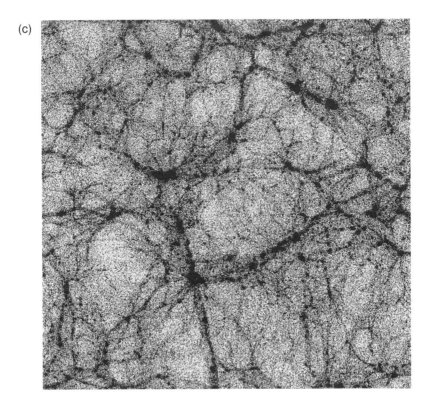

FIGURE 6 These three 'frames' show three stages in the development of clustering in the expanding universe (computed by the Virgo consortium). The scale is adjusted so that each picture shows the same amount of mass (which would, of course, have been more compressed at earlier stages in the expansion). The region shown would be about 300 million light years across.

ing inhomogeneities, and creating progressively steeper temperature gradients – a prerequisite for emergence of the complexity that lies around us ten billion years later, and of which we are part.'

The way cosmic structures evolve is in principle as predictable as the orbits of the planets, which have been understood since Newton's time. But to Newton, some features of the solar sytem were a mystery. He showed why the planets traced out ellipses. It was a mystery to him, however, why they were 'set up' with their orbits almost in the same plane, all circling the sun in the same way. In his *Opticks* he writes:

blind fate could never make all the planets move one and the same way in

> orbits concentrick . . . Such a wonderful uniformity in the planetary system
> must be allowed the effect of choice.

This co-planarity is now understood: it is a natural outcome of the solar system's origin as a spinning protostellar disc.

The demarcation between phenomena that are the manifestations or working out of known laws, and those which are mysterious 'initial conditions' still exists, as sharply as it did for Newton. We are still, at some stage, reduced to saying 'Things are as they are because they were as they were'. The progress has been to push the barrier back from the beginning of the solar system to the first second of the Big Bang.

Just as Newton had to specify the initial trajectories of each planet, our calculations of cosmic structure need to specify a few numbers:

(1) The expansion rate.
(2) The proportions of ordinary atoms (or their constituent quarks), dark matter and radiation.
(3) The character of the fluctuations – large enough to evolve into structures, but not to invalidate the overall uniformity.

Can we take a further step, and explain these numbers in terms of some processes that happened still earlier than the starting point of these simulations?

The trouble is that, the further we extrapolate back, the less confidence we have that known physics is either adequate or applicable. For the first millisecond, everything would have been squeezed denser than an atomic nucleus. For the first 10^{-14} second the energy of every particle would surpass what even the new accelerator at CERN (Centre Européene pour la Recherche Nucléaire) will reach.

Not even the boldest theorists can extrapolate back beyond the stage when quantum effects become important for the entire universe. The two great foundations of twentieth-century physics are, on the one hand, Einstein's theory of gravity (general relativity) and, on the other, the quantum uncertainty principle. But there is generally no overlap between these two great concepts. Gravity is so weak that it is negligible on the scale of single molecules, where quantum effects are crucial. Conversely, gravitating systems such as planets and stars are so large that quantum effects can be ignored in studying how they move. Right back at the beginning of the universe, the densities could have been so high that quantum effects were important for the whole universe. This happens at the Planck time, 10^{-43} second.

It is helpful to divide cosmic history into three parts. Part 1 is the first milli-second, a brief but eventful era spanning forty decades of logarithmic time, starting at the Planck time. This is the intellectual habitat of the high-energy theorist and the quantum cosmologist. Part 2 runs from a millisecond to about a million years. It is an era where cautious empiricists like myself feel more at home. The densities are far below nuclear density, but everything is still expanding in an almost homogeneous fashion. The relevant physics is firmly based on laboratory tests, and theory is corroborated by good quantitative evidence: the cosmic helium abundance, the background radiation, etc. Part 2 of cosmic history, though it lies in the remote past, is the easiest to understand. The tractability lasts only so long as the universe remains amorphous and structureless. When the first gravitationally bound structures condense out – when the first stars, galaxies and quasars have formed and lit up – the era studied by traditional astronomers (Part 3) begins. We then witness complex manifestations of well-known basic laws. Gravity, gas dynamics and feedback effects from early stars combine to initiate the complexities we see around us and of which we are part. Part 3 of cosmic history is difficult for the same reason that all environmental sciences, from meterology to ecology, are difficult.

Then we realize that the few basic numbers that determine how the universe has evolved are all legacies of the uncertain physics of Part 1. I would now like to discuss some ideas about this, and about how even the physical laws themselves may have been imprinted in the ultra-early universe. Be warned that I am now entering speculative territory, where even Zeldovich would harbour some doubts.

First, what about the initial expansion rate? This has to be very precisely tuned. The two eschatologies – perpetual expansion or recollapse to a 'crunch' – seem very different. But our universe is still expanding after ten billion years. Had it recollapsed sooner, there would not have been time for stars to evolve; indeed, if it had collapsed after less than a million years it would have remained opaque, precluding any thermodynamic disequilibrium. On the other hand, the expansion cannot be too much faster than the critical rate. Otherwise gravity would have been overwhelmed by kinetic energy and the clouds that developed into galaxies would have been unable to condense out.

In Newtonian terms the initial potential and kinetic energies were very closely matched. How did this come about? Why does the universe have the large-scale uniformity which is a prerequisite for progress in cosmology?

The answer may lie in something remarkable that happened during the first 10^{-36} second, when our entire observable universe was a few centimetres across. Ever since that time, the cosmic expansion has been *decelerating*, because of the gravitational pull that each part of the universe exerts on everything else. Theoretical physicists have come up with serious (though still, of course, tentative) reasons why, at the colossal densities before that time, a new kind of 'cosmical repulsion' might come into play and overwhelm 'ordinary' gravity. The expansion of the ultra-early universe would then have been exponentially *accelerated*, so that an embryo universe could have inflated, homogenized and established the 'fine-tuned' balance between gravitational and kinetic energy when it was only 10^{-36} second old.

This generic idea that the universe went through a so-called inflationary phase is compellingly attractive. The fluctuations from which clusters and superclusters form, and the even vaster ones whose imprint on the background radiation spreads right across the sky, may be the outcome of microscopic quantum phenomena from an ultra-ancient epoch when the universe was squeezed smaller than a golfball. We do not, of course, know the physics that prevailed at this ultra-early time, but there is a real prospect of discovering something about it. Specific models of how the inflation is driven make distinctive predictions about things we can observe: large-scale clustering, and small non-uniformities in the background radiation over the sky. We shall soon be confronting the inflationary era of cosmic expansion with *real empirical tests*, just as we can already, by measuring the abundances of helium and deuterium, learn about physical conditions during the first few seconds.

Some other remarkable fossils of the ultra-early universe, conjectured by theorists, are being looked for. Among these are magnetic monopoles, and small black holes the size of an atom but weighing as much as a mountain. Even more astonishing are cosmic strings – elastic loops of concentrated energy, thinner than an elementary particle, but long enough to stretch across the universe, flailing around at nearly the speed of light, and heavy enough for their gravity to affect entire galaxies. These would be crucial links between the cosmos and the microworld.

The inflationary idea also, incidentally, strongly suggests that the mean cosmic density is very close to the 'critical' value that demarcates the boundary between perpetual expansion and eventual recollapse. That is the basis of the prejudice I mentioned earlier in favour of the critical density.

The universe has, then, in a sense, zero net energy. Every atom has a rest mass energy: Einstein's mc^2. It also has a negative potential energy due to the gravitational field of everything else, and this exactly balances its rest mass. Thus, it 'costs nothing', as it were, to expand the mass and energy in our universe.

Physicists sometimes loosely express such ideas by saying that the universe can essentially arise 'from nothing'. They should watch their language, especially when talking to philosophers. The physicist's vacuum has all particles and forces latent in it; it is a far richer construct than the philosopher's 'nothing'.

Any such theory would, of course, be hard to check, and may never be taken too seriously unless it has a compelling inevitability about it, a resounding ring of truth that compels assent. In any case it would not tell us *why there was* a universe. To quote Stephen Hawking in *A Brief History of Time* (1988): 'What is it that breathes fire into the questions? Why does the Universe go to all the bother of existing?'

The character of the universe and everything in it depends, of course, on the strengths of the basic physical forces – gravity, electromagnetism, etc. These are also part of our 'initial conditions'.

I already mentioned gravity's counterthermodynamic tendencies. Gravity has a second crucial feature, its feebleness. The gravitational pull between two protons is 36 powers of 10 weaker than the electrical repulsion between them. On large scales, gravity wins because everything has, as it were, the same sign of 'gravitational charge' – there is no cancellation of positive and negative, as in electricity.

Imagine a series of lumps containing successively 10, 100, 1000 atoms and so on. The 24th, containing 10^{24} atoms, would be about the size of a sugar lump. The 40th would be a mountain or small asteroid. The gravitational energy of each atom due to the rest of the lump it belongs to is proportional to the mass/radius (M/R). The radius of a lump is proportional to the cube root of the mass, so gravity gains in importance as the 2/3 power of the number of particles. Gravity is handicapped by 36 powers of 10 to start with. So it becomes competitive only for the 54th lump, containing 10^{54} atoms, because 36 is 2/3 of 54. That lump would be as big as Jupiter. Anything still larger is crushed by gravity, and

155

would become a star. It is because gravity is so weak that stars have to be so massive. In any lesser aggregate, gravity could not squeeze the material to high enough central densities and pressures for nuclear fusion to occur.

Consider now a hypothetical universe where gravity was 10^{10} times stronger than in ours – 'only' 26 rather than 36 powers of 10 weaker than the electrical forces in atoms – but where the microphysics was unchanged. Atoms and molecules would behave just as in the actual universe, but objects would not need to be so large before gravity became competitive. In this imagined universe, mini suns with 10^{-15} times the sun's mass would live for about one year.

The (literally) crushing effect of strong gravity would hamper complex evolution on this hypothetical world. No animals on any planet large enough to retain an atmosphere could be any bigger than insects, and they would need thick legs to support them. More constraining still is the limited time. Chemical and metabolic processes depend on microphysics, and would not be speeded up. But the mini sun would have exhausted its energy before even the first steps in organic evolution had got under way.

If gravity were stronger, there would be fewer powers of 10 between *astrophysical* timescales and the basic microphysical timescales for physical or chemical reactions. Paradoxically, the weaker gravity is – provided it is not exactly zero – the grander and more complex can be its consequences.

Our cosmic environment is exceedingly sensitive to other physical quantities. For instance, if protons had electrical charges just a small percentage greater, then no atoms other than hydrogen could exist: chemistry would be a simple subject. Indeed, if you imagine 'turning a set of knobs' to adjust the physical constants, most choices would lead to 'still-born' universes in which the prevailing laws would not allow *any* complexity to emerge. These universes might, for instance, never deviate from thermodynamic equilibrium, or (perhaps because gravity was very strong) they might exists for too short a time, or (more radically) have only two spatial dimensions. How should we respond to this line of thought? Its implications depend very starkly on what the final theory (if there is one) is like. There are two contrasting possibilities.

One, option A, is that some such final theory fixes all the physical constants uniquely so that they are all calculable from some fundamental equation. The physics governing our universe *could not* then logically have been otherwise. It would then just be a brute fact that the uniquely specified physical constants happened to lie in the narrowly restricted range that allowed such complexity to evolve in the low-energy world we inhabit. The intricate consequences implicit in the fundamental equations may astonish us, but our amazement would be no less subjective than that of a mathematician who is surprised at the intricately interrelated consequences of a simple algorithm. (Take, for instance, the 'Mandelbrot set'. The recipe or algorithm for constructing this astonishing pattern can be written in just a few lines, but it encodes an intricate variety of new structures, however much we magnify it.) Any apparent fine-tuning would have to be accepted as just coincidental.

But there is an alternative, option B. The numbers we call the constants of physics may not be uniquely fixed by the fundamental theory. If an 'ensemble of universes' existed, each one governed by different physics, there would be some in which the conditions were tuned propitiously for complexity to emerge.

This second option would be fulfilled by some variants of inflationary cosmology. According to the Russian cosmologist Andrei Linde, our universe, itself extending far beyond the ten billion light years we can so far see, rather than being 'everything there is' is just one bubble linked to other space–times in an infinite eternally replicating ensemble – the metauniverse. To recall my earlier analogy: the ocean may extend far beyond our horizon, but that does not mean it extends uniformly to infinity.

The physical forces, and the masses of the elementary particles, are the outcome of some kind of phase transition, connected to the force that drives inflation. The imprint left by these phase transitions (the relative strengths of the present-day forces) may then be somewhat arbitrary or 'accidental', like the patterns of ice on a pond, or the way a magnet behaves when cooled. The 'constants of nature' would have quite different values in other universes in the ensemble.

If this (very schematic!) picture had anything in it, the apparent 'fine-tuning' need not surprise us at all. The fundamental constants would be the outcome of random accidents (option B), so in a sufficiently capacious metauniverse the physics would inevitably turn out propitious for complexity to evolve in some members of the ensemble, and we are obviously not surprised to find ourselves in one of these.

This strays dangerously close to what is called anthropic reasoning. Fortunately, there is no space here to stray further, but I do not think anthropic reasoning is quite as silly or vacuous as it is sometimes made to sound. Indeed, it gives us extra grounds for suspecting that any final theory will have the permissive character of option B: there is then nothing surprising about the existence of a universe with specific 'coincidental' features.

I earlier gave an implicit 'health warning', urging that you should not necessarily believe all the cosmology you read in the newspapers. Let me conclude by trying to assess the state of play – where we can lay confident bets and where we should not (or not yet).

Cosmology has made quite amazing progress since the years when the now-abandoned steady-state theory was being boisterously debated, especially here in Cambridge. That theory, then, seemed specially attractive because, if it were correct, every evolutionary process (from atoms to galaxies) had to be going on somewhere now, and so should be observable and could be investigated. At that time, it was felt that a Big Bang theory could never be made scientific, because the key processes would be shrouded in the remote past. They are, indeed, in the remote past, but what is remarkable is that they are not inaccessible to study. Telescopes can view directly 90% of cosmic history; other techniques can probe still earlier phases. We are confident about cosmic history, at least in broad outline, back to one second, when the first elements were made – back to the start of what I have called Part 2 of cosmic history. The challenge is now to delineate more fully how an almost featureless fireball evolved into the cosmos of which we are a part ten billion years later.

The last illustration (Figure 7) shows Einstein: the young Einstein, not the benign and unkempt sage of poster and **T**-shirt. One of his best-known remarks is that 'The most incomprehensible thing about the universe is that it is comprehensible'. Cosmology has progressed because the laws of physics we study in the laboratory apply in the remotest quasar, and back to the first few seconds of the 'Big Bang'. When there is no firm link with laboratory science, cosmological inferences are more fragile. That is why we are on shakier ground when we venture back into the first millisecond, and we should not disguise this. Our methodology is no longer like that of a geologist, or practitioner of other historical sciences: new basic physics has to be discovered, rather than established physics being applied.

Ideas about the ultra-early universe are often presented in popular books in the same tone as descriptions of the Hubble Law, the microwave background, etc. This makes me somewhat uneasy: there is a risk that overcredulous readers may accept tentative speculations; on the other hand, more sceptical readers

FIGURE 7 The young Einstein.

may overlook the strengthening range of observations that buttress claims about the later stages of cosmic evolution.

Some previously speculative questions are now coming within the scope of serious science. In the ultra-early universe, the mysteries of the cosmos and the microworld overlap. Processes as early as 10^{-36} second may have imprinted the excess of matter over anti-matter, the ripples in the fabric of space–time, and perhaps the physical laws themselves.

Modern cosmology has been moulded by cultural climate, and given an

159

impetus by the influx of scientists – particle physicists, for instance – with different expertise and style. It has been moulded further by the opportunities and constraints of the available techniques: experimental, observational and computational. These sociological dimensions are in themselves fascinating. However, such studies should not obscure what to us 'in the zoo' seems the essence of our science: that it is a collective and cumulative enterprise which, albeit fitfully, is bringing the workings of the cosmos into a sharper and 'truer' focus.

FURTHER READING

Audouze, J. and Israel, G. (eds.) *The Cambridge Atlas of Astronomy*, 3rd edition. Cambridge: Cambridge University Press, 1994.

Barrow, J. *The Origin of the Universe*, London: Weidenfeld & Nicolson, 1995.

Begelman, M. and Rees, M. *Gravity's Fatal Attraction: Black Holes in the Universe*, New York: W. H. Freeman, 1995.

Rees, M. *Perspectives in Astrophysical Cosmology*, Cambridge: Cambridge University Press, 1995.

Rees, M. *Before the Beginning: Our Universe and Others*, London: Simon & Schuster; New York: Addison Wesley, 1997.

Silk, J. *A Short History of the Universe*, New York: W. H. Freeman, 1995.

Notes on Contributors

Gillian Beer is King Edward VII Professor of English and President of Clare Hall at the University of Cambridge. Her books *Darwin's Plots* (1983) and *Open Fields: Science in Cultural Encounter* (1997) engage with questions raised in this essay. She has also written widely on fiction and narrative.

Jared Diamond's interests range from evolutionary physiology, studied in the laboratory, to birds studied in the jungles of New Guinea. His book *The Rise and Fall of the Third Chimpanzee*, a passionate account of human evolution, won Britain's Rhône-Poulenc Science Book Prize for 1992. His just-published book *Guns, Germs, and Steel* compares the development of human societies on all the continents for the last 13 000 years. He has been elected to the National Academy of Sciences, the American Academy of Arts and Sciences, and the American Philosophical Society.

Freeman Dyson, FRS, is Emeritus Professor of Physics at the Institute for Advanced Study in Princeton. His research work is scattered over many fields, from pure mathematics to particle physics and nuclear engineering. His most important contribution was to clarify quantum electrodynamics, the physical theory that describes the interactions between matter and electromagnetic fields. He has written a number of books about science for the general public, including *Disturbing the Universe, Origins of Life* and *From Eros to Gaia*.

Andrew Fabian, FRS, is a Royal Society Research Professor at the University of Cambridge, and a Fellow of Darwin College. He researches in astronomy with main interests in active galaxies and clusters of galaxies. He organized the first Darwin College Lecture series, on Origins.

Stephen Jay Gould is the Alexander Agassiz Professor of Zoology and Professor of Geology at Harvard University. He is also curator of invertebrate palaeontology at Harvard's Museum of Comparative Zoology. He has

161

written extensively on aspects of evolutionary science, in both professional and popular books.

Tim Ingold is Max Gluckman Professor of Social Anthropology at the University of Manchester. He has carried out ethnographic research in Finnish Lapland, and has written extensively on comparative questions of environment, technology and social organization in the circumpolar North, as well as on evolutionary theory in anthropology, biology and history, on the role of animals in human society and on issues in human ecology. He is the author of *Evolution and Social Life* (1986) and *The Appropriation of Nature* (1986). He is currently working on aspects of the anthropology of technology and environmental perception.

Martin Rees, FRS, is a Royal Society Research Professor at the University of Cambridge, and a Fellow of King's College. He was until 1992 Plumian Professor of Astronomy and Director of the Institute of Astronomy at Cambridge, and before that a Professor at Sussex University. Sir Martin is the Astronomer Royal.

Richard Rogers was educated at the Architectural Association, with post-graduate training at Yale University. He founded Team 4 with Sue Rogers and Norman and Wendy Foster in 1963 and then a partnership with Renzo Piano in 1971. He founded the Richard Rogers Partnership in 1977. Lord Rogers was awarded the Chevalier, L'Ordre National de la Légion d'Honneur in 1986, was knighted in 1991 and made a Life Baron (The Lord Rogers of Riverside) in 1996.

Lewis Wolpert, FRS, is Professor of Biology as Applied to Medicine, at University College London. His main interest is pattern formation in embryonic development. Among other works, he has written *Triumph of the Embryo* (1991).

Acknowledgements

CHAPTER 1

FIGURE 1 DAR225/120. By Leonard Darwin (1878). By permission of the Syndics of
Cambridge University Library.

FIGURE 2 By permission of the President and Council of the Royal Society.

CHAPTER 2

FIGURES 1, 3, 4 and 5 Reproduced with permission from Wolpert, L. (ed.) *Principles
of Development*, London: Current Biology Ltd, 1997.

FIGURE 2 Reproduced with permission from Thompson, D'Arcy W. *On Growth and
Form*, new edition, Cambridge: Cambridge University Press, 1942.

FIGURES 6 and 7 Reproduced with permission of the Company of Biologists Ltd
from Akam, M, Holland, P., Ingham, P. and Wray, G. (eds.) 'The evolution of
developmental mechanisms', *Development Supplement* 1994.

FIGURE 8 Reproduced with permission from the Company of Biologists Ltd from
Wolpert, L. *Development Supplement* 7 (1992), 7–13.

CHAPTER 3

FIGURE 1 Reproduced with permission from Diamond, J. M. *Guns, Germs, and Steel*,
Fig. 10.1, New York: W. W. Norton; London: Jonathan Cape/Random House,
1997.

CHAPTER 4

FIGURE 1 Reproduced with permission of Cambridge University Collection of Air
photographs: copyright reserved.

CHAPTER 5

FIGURE 2 Adapted from Ingold, T. ' "People like us": the concept of the
anatomically modern human', *Cultural Dynamics* 7 (1995), 211, Fig. 5, by
permission of Sage Publications Ltd.

CHAPTER 8

FIGURE 1 Courtesy of Dr Christopher Burrows, ESA/STScI and NASA. UK HST
Support Facility.

FIGURE 2 Courtesy of Jeff Hester and Paul Scowen, Arizona State University and NASA. UK HST Support Facility.

FIGURE 3 Courtesy of Robert Williams and the Hubble Deep Field Team (STScI) and NASA. UK HST Support Facility.

FIGURE 4 Reproduced with permission from *The Independent*, 24 April 1992, Susan Watts and Tom Wilkie.

FIGURE 5 Courtesy of W. Couch, University of New South Wales, R. Ellis, Cambridge University, and NASA.

FIGURE 7 Permission granted by the Albert Einstein Archives, the Jewish National and University Library, the Hebrew University of Jerusalem, Israel.

Index

Numbers in italics indicate figures.